the brain

a beginner's guide

From anarchism to artificial intelligence and genetics to global terrorism, Beginners' Guides equip readers with the tools to fully understand the most challenging and important debates of our age. Written by experts in a clear and accessible style, books in this series are substantial enough to be thorough but compact enough to be read by anyone wanting to know more about the world they live in.

Other titles available in this series:

anarchism
ruth kinna

anti-capitalism
simon tormey

artificial intelligence
blay whitby

bioterror and biowarfare
malcolm dando

democracy
david beetham

energy
vaclav smil

evolution
burton guttman

evolutionary psychology
louise barrett, robin dunbar, john lycett

genetics
a. griffiths, b. guttman, d. suzuki, t. cullis

global terrorism
leonard weinberg

nato
jennifer medcalf

the palestine–israeli conflict
dan cohn-sherbok, dawoud el-alami

postmodernism
kevin hart

quantum physics
alastair i. m. rae

religion
martin forward

Forthcoming:

the arms trade
matthew schroeder, rachel stohl & daniel m. smith

biodiversity
john spicer

capitalism
andrew kilmister, gary browning

criminal psychology
ray bull

the brain

a beginner's guide

ammar al-chalabi, martin r. turner
and r. shane delamont

ONEWORLD

OXFORD

the brain: a beginner's guide

Oneworld Publications
(Sales and Editorial)
185 Banbury Road
Oxford OX2 7AR
England
www.oneworld-publications.com

ISBN-13: 978–1–85168–373–4
ISBN-10: 1–85168–373–9

Cover design by the Bridgewater Book Company
Typeset by SNP Best-set Typesetter Ltd., Hong Kong
Printed and bound by WS Bookwell, Finland

Colour images reproduced with permission: plates 1–4 courtesy of Dr
Catherine Richards, Glenfield General Hospital; plates 5–12 courtesy of
Professor Paul Matthews, Oxford Centre for Functional Magnetic Resonance
Imaging of the Brain (FMRIB) (www.fmrib.ox.ac.uk/image_gallery); plates 13
and 14 courtesy of Dr Sharon Abrahams, Department of Human Cognitive
Neuroscience-Psychology, The University of Edinburgh; plates 15 and 16
courtesy of the Neuroimaging Research Group, Centre for Neuroimaging
Sciences at the Institute of Psychiatry.

contents

appendices 175

list of plates

Plate 1. Sagittal section through the brain.
Plate 2. Undersurface of the brain.
Plate 3. Close up undersurface of the brain.
Plate 4. Coronal section through the brain.
Plate 5. fMRI scan showing the brain regions activated by button pressing.
Plate 6. fMRI scan showing the brain regions activated by moving an arm.
Plates 7, 8 and 9. fMRI scans showing brain areas activated in someone experiencing simultaneous audio and visual stimulation.
Plate 10. fMRI scan showing frontal activation.
Plate 11. fMRI scan showing the cerebellum during activity involving hand–eye coordination.
Plate 12. fMRI scan showing hand function following a stroke.
Plate 13. fMRI scan showing brain activity when subjects are asked to list words beginning with a particular letter.
Plate 14. fMRI scan showing brain activity when subjects are shown a picture of an animal and asked to name it.
Plate 15. A 3D image reconstructed from MRI scans showing the corticospinal tract.
Plate 16. A 3D MRI-generated image showing the organization of individual nerve tracts in the internal capsule.

preface

When we were first approached to write this book, it was with a mixture of excitement and trepidation that we said yes: excitement because the brain is the essence of who we are and trepidation because the brain is the least understood of all organs. This book is aimed at the intelligent, educated reader. We have tried to keep the jargon as low as possible, but the content remains challenging. The emphasis is on a scientific perspective: we aim to lead the reader from a state of assumed ignorance to an understanding of cells, neurons, brains and brain functions, such as consciousness, explaining the development of the nervous system and the development of the mind along the way. We have tried to keep it as up to date as possible, but the 1990s were the *Decade of the Brain*, with huge investment in neuroscience. As a result, there has been an explosion in research and it is possible that some of the concepts in the book will be overturned in the next few years. Although we have kept to facts for the basics, for some aspects of the brain we can only provide opinions, because few facts exist, but these opinions are based on our existing knowledge. We are all neuroscientists and neurologists dealing with the brain every day and some of the ideas in this book are therefore our own original thoughts. As far as possible, we have tried to explain parallel concepts needed to understand the workings of the brain, especially where these are not related to neuroscience directly. It has been a wonderful experience putting our thoughts on paper, and we hope you enjoy reading this book as much as we enjoyed writing it.

Ammar Al-Chalabi, Martin R. Turner and R. Shane Delamont, December 2004.

acknowledgements

Many people have been instrumental in making this book possible. We want to thank our kind and very patient editor, Victoria Roddam at Oneworld, our all knowing and highly skilled copyeditor, Ann Grand, and the unknown test readers whose suggestions have greatly improved the manuscript. Many friends and family have helped immeasurably by reading or listening patiently to different versions of the text, often many times over. In particular, this book would not have been possible without the support of Cathy Richards, Sally Coutts and Sue Richards.

Illustrations throughout the book are by Charlotte James, our excellent and sharp artist.

Chapter contributions were: overall editing for style, AAC; introduction AAC; chapter 1, The history of the human brain MRT; chapter 2, The evolution of the brain AAC; chapter 3, Nerves AAC; chapter 4, The development of the brain AAC; chapter 5, The anatomy of the brain MRT; chapter 6, The supporting structures of the brain MRT; chapter 7, The development of behaviour and reasoning AAC; chapter 8, Consciousness AAC; chapter 9, Memory RSD and AAC; chapter 10, Sleep RSD and AAC; chapter 11, The motor system AAC; chapter 12, The sensory system AAC and RSD, chapter 13, The visuospatial system MRT; chapter 14, Language, hearing and music AAC; chapter 15, Emotions and the limbic system AAC and RSD; chapter 16, Investigating the brain MRT; chapter 17, Living for ever AAC; chapter 18, The end? AAC.

introducing the brain

in the beginning . . .

If the brain were so simple we could understand it, we would be so simple we couldn't.

Lyall Watson (contemporary author)

Thought and intelligence have always been of great importance to humankind and are regarded as the essence of existence: "I think therefore I am," wrote Descartes in 1641. Yet it is only relatively recently that we have regarded the brain as the source of thoughts, reason, emotion and being. It is now so closely identified with our selves that in many countries brain death is regarded as the same as actual death.

In the great stories of creation, the first thing in existence, before there was even light, was an intelligent mind, because the creation of the universe required a being intelligent enough to design and make it. In the modern scientific version of events, we believe that one of the last things to come into being was an intelligent mind, because something so complex could only arise after many millions of years of evolution. In both cases, intelligence (and by implication the brain) has a special place.

The brain is the most complex object known. It is a chemical and electrical powerhouse, sending messages where needed, in a perfectly targeted way. This soft, grey, one and a half kilogram organ is not only where we experience and manipulate the world but is also responsible for control of our breathing, body temperature, blood pressure and hormones. Billions of nerve cells, each specifically connected to thousands of others, are its essence. Keeping these cells alive and well for a lifetime requires an enormous amount of sophisticated help from the rest of the body and an integrated life support system. To protect the whole from the outside world it is wrapped in a lining, bathed in shock-absorbing fluid and packaged in a strong bony box.

But this description does not answer the simplest question: why do we need a brain? Plenty of living things survive perfectly well without a brain and plenty more survive with what can only be described as a trivial collection of nerves. In this book we will try to answer this question and, along the way, explain what the brain is made of, how it is put together, what it does and how it does it.

Modern medicine divides problems with the brain into two groups: neurology and psychiatry, corresponding to problems with the nervous system and problems with the mind. These disciplines attempt to explain disturbances in experience, behaviour, sensation, movement or speech and relate them to physical, chemical or electrical disturbances in the relevant area of the nervous system. Our ancestors did not make such associations, for example believing that seizures were caused by demonic possession, or that hearing voices might be a sign that God was speaking. Many people also believed, and still believe, in another ability of the brain: psychic or paranormal power. These ways of thinking about the brain are important; part of the natural tendency we have to categorize and try to make sense of the world around us. Even modern physics recognizes that simple observation by a conscious being affects the behaviour of the subatomic world; indeed this is a corner-stone of quantum theory. This brings us to the thorny question of consciousness. What is it? What makes it? Are we conscious when we sleep? Is there consciousness after death? These questions have puzzled people for centuries and we will address them here, with an attempt at a modern answer.

Our imaginative brain is what separates us, in our own minds, from other animals. Research is now beginning to show that we are perhaps not so different after all. In this book we want to show that human brains are special and amazing and that although we know an enormous amount about them, there remains far more still to learn. We will start with a look at the past and how people came to realize that the brain is an important organ.

the history of the human brain – so it does do something after all

The brain is a world consisting of a number of unexplored continents and great stretches of unknown territory.

Santiago Ramon y Cajal (Spanish physician and anatomist, 1852–1934)

I believe in an open mind, but not so open that your brains fall out.

Arthur Hays Sulzberger (US newspaper publisher, 1891–1968)

early beliefs – before history began

The brain has no moving parts: unlike the heart, lungs or intestines it does not pulsate, inflate or squeeze. The brain does not make anything: unlike the kidneys, liver or spleen, no urine, bile or lymph comes out of it. Unlike the skin or bones, the brain serves no obvious purpose and yet we now believe it is responsible for thought, emotion and free will. How did we come to such a conclusion and what did people think before? To answer this, we must travel to the past and step from conclusion to conclusion to the present day. We begin this tour with three cautions. First, although our ideas about thinking and emotion stretch back to the earliest recorded civilizations, our knowledge of these early beliefs is based on archaeological evidence and is therefore very patchy. Second, the history of medicine is hugely biased towards Western historical documentation and so we will inevitably mainly describe this view of things. Finally, attitudes to animal and human experiments were quite different in the past and many of the experiments we will describe are quite unpleasant and would be highly unlikely to be allowed today.

Olympic thinkers – the Greeks and their legacy

> Never trust anything that can think for itself if you can't see where it keeps its brain.
>
> J. K. Rowling (*Harry Potter and the Chamber of Secrets*)

The simple view of the brain as the most fundamental of all organs may seem rather obvious, but even this assumption is based on knowledge largely acquired in the last 200 years. Prior to the Greek philosophers, the heart was widely held to be the seat of intellect. Indeed, a scholar from ancient Egypt, Herodotus (485–425 BCE), writing about mummification, documented the great care taken in the preparation of organs such as the heart, lungs, liver, stomach and intestines, whilst the brain was simply scooped from the skull. The Ancient Egyptians saw the number of visible connections running to and from the heart as evidence for its importance, whereas the brain did not seem to do much. Nevertheless, it was around this time that the first documented ideas about the true function of the brain were recorded, not by the Egyptians, but by the Ancient Greeks.

Opinions on thinking and emotion were, for a time, dominated by three major philosophers but only two of these thought that the brain was important. Hippocrates (460–370 BCE), the "Father of Medicine", wrote in his book *The Sacred Disease*,

> Men ought to know that from the human brain and from the brain only arise our pleasures, joys, laughter, and jests as well as our sorrows, pains, griefs and tears . . . It is the same thing which makes us mad or delirious, inspires us with dread and fear, whether by night or by day, brings us sleeplessness, inopportune mistakes, aimless anxieties, absent-mindedness and acts that are contrary to habit.

He also noted a fundamental property of the brain's wiring: damage to the brain on one side produced a bodily deficit on the opposite side.

The philosopher Plato (428–347 BCE) proposed that the "vital principle" lay within the brain, which in conjunction with the spinal cord was responsible for the control of "vital force". To quote a translation of his work: "Copying the round shape of the universe, they (*the gods*) confined the two divine revolutions (*the eyes*) in a spherical body – the head, as we now call it – which is the most divine part of us and lords over all the rest."

Aristotle (384–322 BCE), however, believed that the function of the brain was to "cool the heart", although he did also draw the conclusion that the size of this "cooling apparatus" might be linked to overall intelligence. His heart-centred theory was based on his observation that in the embryo the heart is the first organ to develop and is also warmer in temperature, which he felt was a direct measure of an organ's involvement in vital processes. He also noticed that a chicken exhibited a life of its own, running around after the head was removed, which was further evidence that the heart was needed for action rather than the brain.

A series of Ancient Greek physicians gradually took us, over a number of years, to a more brain-centred view. The first, Strato (340–290 BCE), refined Plato's original localization of the "vital principle" to the frontal region between the eyebrows, the second, Xenocrates (396–314 BCE), to the crown of the head. The third, Herophilus (335–280 BCE), carried out extensive dissection of the human body and recognized the brain, particularly its base, as the centre of the nervous system, even noticing a difference between nerves for sensation and

"MIKE THE HEADLESS CHICKEN FOR PRESIDENT!"

A rather gruesome example of a chicken surviving without a head makes it easier for us to understand exactly why the heart was so long considered the source of thoughts and emotion while the brain was not. "Mike the Headless Chicken" (http://www.miketheheadless-chicken.org) lived on for eighteen months after a failed attempt to kill him. In 1945, he was a rooster destined for his owners' dinner table. After having his head cut off with an axe, his body ran around, as is usual for chickens. But rather than eventually stopping and dying, he instead "returned to the job of being a normal chicken" and began preening and pecking with his neck. He was fed through an eye dropper until dying one night, apparently from choking. He probably survived because, in a chicken, many reflexes are stored in the spinal cord and brainstem, and the blow was high enough to leave a significant proportion intact. It is thought that he developed a blood clot at the time of the butchery, preventing his death from bleeding. More likely, the blow severed the arteries in such a way that the muscular walls sealed up automatically, as they are designed to. Mike's spirit is celebrated every May in his home town of Fruita, Colorado.

nerves for action. His discovery of the fluid cavities within the brain, the ventricles, provided the basis for the later "ventricular doctrine" of brain function. Finally, the Alexandrian physician Erasistratus (304–250 BCE) suggested that the greater intelligence of humans might be attributed to the greater number of folds or wrinkles in their brains, compared with those of other animals. However, although more brain-centred ideas were beginning to develop, Aristotle's view of the brain as a glorified air-conditioner persisted in many circles until medieval times.

Over several hundred years, from the third century, one anatomist dominated all thinking: the Greek physician and philosopher Galen (130–200 CE). By dissecting animals (but not humans), he developed anatomical ideas which were taught in all medical schools. His concept of a "physiology of spirits" described a vital force called "pneuma" that, mixed with blood, travelled to the brain, where it was given "animal spirit". This then controlled the brain, nerves and feelings. The animal spirit was stored in the ventricles (a set of fluid-filled cavities within the brain) and sent through hollow nerves to produce movement and sensation. The fourth century theologian and bishop of Emesa, Nemesius (b. 320 CE), in his work *On the Nature of Man*, further developed this proposal with the so-called "ventricular doctrine": the idea that the key elements of imagination, intellect and memory were localized to the ventricular system.

More than a millennium later, in 1543, the Renaissance anatomist, Andreas Vesalius (1514–1564), wrote a detailed anatomical atlas, based on the dissection of human corpses, that challenged Galen's ideas and forever changed the way anatomy was taught in the west. It was called *De Humanis Corpora Fabrica* (*On the Fabric of the Human Body*) and described the five ventricles of the human brain (we now number four of them and name the fifth – see glossary). Vesalius also ascribed three souls to people and assigned to the brain "the chief soul, the sum of the animal spirits, whose functions are distinctly mental". He was also first to discern the difference between the grey and white matter of the brain. His description referred to the greyish appearance of the thin rim around the main substance of the brain, termed the cortex, contrasted with the whiter-appearing bulk of the brain tissue, its appearance, as we now know, due to the insulation wrapped around the nerves.

lumps and bumps – the art of phrenology

The first serious attempts at localization of brain function began with the development of phrenology, by the renowned Viennese neuro-anatomist Franz Joseph Gall (1758–1828). This doctrine, described in

Gall's work *The Anatomy and Physiology of the Nervous System in General, and of the Brain in Particular*, saw the excellence of mental faculties or traits as being determined by the size of the brain area upon which they depended, an idea that to some extent we would agree with today. In turn, he thought, the size of these brain areas could be judged by the development of the skull and the bumps overlying each area, an idea we now consider ridiculous.

Gall, with his colleague Johann Spurzheim (1776–1832), identified thirty-seven "mental and moral faculties" which they thought were represented on the exterior surface of the skull. These faculties were divided into several spheres, such as intellect, perceptiveness, mental energy and love. Most of the faculties dealt with abstract personality traits such as firmness, cautiousness, marvellousness and spirituality. A chart of the skull was developed, mapping the regions where the bumps and depressions related to these traits could be palpated, measured and diagnosed. White porcelain heads with these maps drawn on them are ubiquitous in antique shops around the world.

Phrenology was widely taken up in general practice. Inevitably, however, it was challenged, particularly by the French scientist Georges Cuvier (1769–1832). Gall himself was hounded out of Vienna by religious and political forces, only to settle in France. It is said that one of the final nails in the coffin of phrenology came when Gall's interpretation of Napolean Bonaparte's skull failed to recognize all the noble qualities the French dictator possessed. Phrenology had all but died as a generally accepted concept by the end of the nineteenth century, though the British psychiatrist Bernard Hollander (1864–1934) persisted with the idea and the British Phrenological Society was listed until the late 1960s. Indeed phrases still used today, such as "you need your head read" and "high-brow" have their basis in phrenology.

illuminating times – the nineteenth century scientists

This brings us to the nineteenth century and a hyperbolic increase in knowledge. For a physician, to have been alive during these times must have been extremely exciting. There were many players but only a few names have withstood the test of time.

Marie Jean Pierre Flourens (1794–1867), a French physiologist, began experiments to investigate the validity of Gall's ideas. He selectively destroyed various parts of the brains of animals and stimulated both animal and human brains with electricity. He also carried out post-mortem studies of the brains of patients with significant mental or neurological deficits. When Flourens removed the two main halves, or hemispheres, of the brains of animals, he noted that all "perceptions and judgement" were abolished. This led him to the correct conclusion that the cerebral hemispheres contained the higher cognitive functions. He also removed the small, separate, ridged structure at the back of and below the cerebral hemispheres – the cerebellum (see chapter 5 – Anatomy) – which resulted in the loss of the animal's co-ordination. Finally, noting death as a result of destruction of the lower part of the brainstem (termed the *medulla oblongata*, which emerges at the base of the brain and is connected to the spinal cord), he deduced that vital functions such as breathing and circulation were regulated there.

Flourens's use of small animal brains did not provide information on the detailed localization of human functions but it was becoming possible to electrically stimulate the brain precisely and experiments could now involve the larger brains of primates and dogs. Enter Paul Pierre Broca (1824–1880), a French surgeon and anthropologist of enormous intellect, whose classic experiments with patients with severe deficits of speech led to the localization of language to a region on the left side towards the front of the brain. One of his patients could only say the word "tan" (and was called Tan by the staff as a result). After his death, Broca discovered a small area of the left side of his brain had been destroyed by syphilis. This region is now called Broca's area.

A British neurologist, John Hughlings Jackson (1835–1911), later developed the idea of Broca's area as the seat of language output, or expression. More or less simultaneously, the German neurologist Carl Wernicke (1848–1904) found a related area further back on the left, concerned with the understanding of language and, moreover, connected to Broca's area by nerve fibre pathways. Thus the basis for our modern and even now still evolving, model of language was born (see chapter 14 – Language, hearing and music).

Two German physiologists, Gustav Fritsch (1838–1927) and Eduard Hitzig (1838–1907), began, through electrical stimulation of dog brains, to map the various points in the motor cortex of the brain by observing different limbs twitching in response to different sites of stimulation. These techniques were refined and greatly developed by the British neurologist and physiologist Sir David Ferrier (1843–1928), who also removed areas of the brain that he stimulated to demonstrate loss of specific movement functions. He brought all his ideas together in 1876 in his publication, *The Functions of the Brain*.

These and other experiments mean that we now have a reasonable understanding of many brain functions and the parts of the brain responsible for them, but it is less clear why we have developed such a large brain in the first place.

the evolution of the brain – how the brain came to be

The most important scientific revolutions all include, as their only common feature, the dethronement of human arrogance from one pedestal after another of previous convictions about our centrality in the cosmos.

Stephen Jay Gould (biologist, 1941–2002)

truth, faith and the way things are

Much has been written about evolution and, surprisingly perhaps, of all modern scientific theories, it alone remains controversial with some, even though the evidence is overwhelming. Dismissing evolutionary theory is a little like dismissing atomic theory or the theory of relativity and while a few people may truly have the knowledge to do so, for many it is a matter of faith. The subject has been discussed many times before and rather than trying to justify it as the correct way of thinking about how we came to be, in this book we will simply accept the process of evolution as the driving force behind the development of the human brain. If you do not feel able to accept this, it will not affect the concepts in the rest of the book, but you will need to find your own explanation for why our brain is as it is.

evolution – the iteration of perfection

Evolution is a process of change between generations, each change usually small in its own right, but accumulating over time, so that when compared with the original, a large change is apparent. A simple example to think about is language. Afrikaans, Dutch and German are all, clearly, related languages. They are thought to have derived from a common German-like ancestor, but how could this happen? The original speakers of this common tongue could understand their children. No one would argue that their children spoke a different language, yet children often use a few words that their parents do not, such as slang. They may also use words differently or use phrases that parents do not understand. In the case of Afrikaans, Dutch and German, over many generations the differences became great enough so that when compared with the original and with each other, they could be thought of as different languages. There was never a time when a new language was created, just an eventual difference between the start and end points. Because of geographical isolation, new words, accents or phrases from one region could not easily pass into common use in another, and gradually the differences became great enough for us to say they were different languages.

This process has three important properties. First, the change is gradual, with each generation being able to understand and be understood by those immediately above and below, so at any one stage it is imperceptible. Second, the change is not planned, designed or deliberate. It just happens as part of the transmission of language from one generation to the next, because the language in the children is not a perfect copy of the language of the parents. Third, because of geographical isolation, common words or phrases used by one group of speakers do not necessarily pass to another, leading to separate evolutionary paths. Humans need to classify – and enjoy it. Thinking of the different points in this continuous process as discrete languages is natural. This process of gradual change is called evolution and will occur in any system which involves the copying of information. There is also a fourth, related, property of this process: selection. Some words or phrases catch on and become popular, whilst others do not. The popular words are used by more people and therefore persist through the generations, whilst the less popular ones die out. In other words, some words are selected by the speakers of the language, while others are not. Again, the process of selection is passive. Nobody *decides* which words will be popular. Some just fit into the linguistic environment of the time better than others.

the evolution of the brain – gradually achieving complexity

A similar process of continuous change and selection has gone on in all aspects of life on earth, including the human brain. Just as languages are the product of sounds and grammatical rules copied into the speech regions of an offspring's brain, organisms are the product of genes and the genetic code copied into an offspring cell, which, for our purposes, results not only in a full organism, but also in its brain. In brains, more complicated wiring schemes provided an advantage in some situations, so they were sometimes selected. The primitive nerve-like cell gave rise to more sophisticated versions and to organisms with nerve networks, then ganglia and finally the complex collection of billions of nerve cells that we call a brain. At no point did a parent give birth to a new species but the small differences slowly added up. Because we can only see a snapshot of a continuous process, we see different species with their particular brains. This is a little like taking a cross-section through the branches of a tree. We would see apparently unconnected circles of wood, which are in reality part of the same structure. Wood circles which appear similar or close together might be classified as related but without knowing the full shape of the tree it is impossible to know with certainty.

The human brain seems, at first glance, to be a mushy disorganised mess of nerves, but it is far from this. There are more potential inter-connections between nerves in the human brain than there are atoms in the known universe. It has a predictable and highly organized struc-ture, which is a product of its past. To understand the chain of events that has led to this structure, we need to accept some concepts that derive from the passive processes of evolution and selection. One is that complexity only arises from simplicity (although this does not mean that simplicity cannot arise from complexity). For example, we think that mammals must have had distant ancestors that were simpler, and that their ancestors were simpler still, all the way back to the origins of life. This is in stark contrast to the basic concepts of creationism and intelligent design in which the ultimate complexity, the omnipotent God, precedes the simpler forms He creates and in which it is accept-able for any level of complexity to be spontaneously created.

Another consequence of evolution is that, during the development of an organism, it echoes its evolutionary past. For example, human embryos have gills, gill arches and a tail during their development; indeed, the embryos of mammals are virtually indistinguishable. This is what we would expect if all mammals had a common ancestor, because this would echo a common evolutionary past. Also, the embryos of birds and mammals are extremely similar to each other in early stages of development, but become less so as the organisms grow. Again, this is what we would expect, because both groups are land-dwelling vertebrates and therefore had a distant common ancestor.

A third consequence is that similarities will be seen in the natural world, because we carry our evolutionary baggage around with us. Just as Afrikaans, Dutch and German all have similar words for "thank you", all modern organisms share the same cell designs, same cell machinery and the same proteins. We share ninety per cent of our genes with yeast. Hundreds of thousands of similar examples strongly argue for a common beginning to living things.

In the natural world there is another reason for similarities. Sometimes they are the only solution to the evolutionary problem (the species have undergone parallel evolution). At their most basic these are features such as symmetry, having a head end, or sensing the environment, but we can see this effect all the way down to the molecular level. At their most striking, these are events like the marsupial and placental wolves, both of which have developed the "wolf" solution to their environment independently. The linguistic equivalent of this is probably the word "mama", which is one of the first sounds a baby can make. The fact all languages have a similar word is not necessarily because they all derived from a single language, but because they all came up with the same word independently for the same reason.

why do we need a brain?

From this discussion, we are forced to conclude that all animals that have a brain must have a shared reason for needing a brain, and that all animals that have a brain like ours must share a common ancestor with us. We will discuss the reason for needing a brain in more detail later, but for now it is sufficient to notice that all organisms that possess a brain share the ability to move from place to place.

start at the beginning

The more complex something is, the more time it would have needed to have evolved. Our brains are highly complex and have evolved over a long period of time. In addition, because each generation is a refinement of the version before, not a brand new design, some of what we see can seem illogical (see box on "A creationist's puzzle"). To under-

A CREATIONIST'S PUZZLE

We carry with us all the evolutionary baggage of the forms that preceded us. One excellent example is the recurrent laryngeal nerve. This nerve is responsible for moving the vocal cords in the larynx, allowing us to speak. If we were designed from scratch by a creator, we should expect the recurrent laryngeal nerve to leave the brainstem and travel the short distance to the larynx directly, because this would be the most logical design. Interestingly, it does not. Instead, the left recurrent laryngeal nerve leaves the brainstem, travels down the neck, loops round the aorta (the large artery which leaves the heart) and travels back up the neck to the larynx. Because it runs back on itself, it is "recurrent". In a human, this is a journey of more than half a metre when about ten centimetres would do; in a giraffe, a journey of about seven metres when a few centimetres would do. This is a bit like sending the control wire for the joystick in a plane all the way to the tail of the plane and back for nothing. A creationist can only easily explain this by saying it is one of God's mysteries, or that God did it deliberately for some unknown reason. For an evolutionist, the explanation is simpler: in the first common ancestor with a recurrent laryngeal nerve, this must have been the simplest solution for connecting the nerve to the larynx. Amphibians today are similar to those in the fossil record and can therefore be regarded as similar to simpler precursors of modern vertebrates. In a frog, the path of the recurrent laryngeal nerve makes sense because the nerve, arch of the aorta and larynx all lie on a straight line, so the route is the quickest and most logical. In reptiles, a slightly more recent evolutionary development, the heart is placed a little lower in the chest and the path for the recurrent laryngeal nerve makes a gentle curve and only a small detour. In mammals, which evolved from a common ancestor with reptiles even more recently, the heart is placed deep in the chest and the larynx well up in the neck. The detour is therefore long, but logical, once we understand that it is a refinement of previous designs.

stand how our brain came to be the way it is, we need to start at the beginning, with a simple, general blueprint, brain: that of a fish.

the brains of a fish

The basic brain plan of the fish nervous system is a tube with three swellings at one end, corresponding to the forebrain, midbrain and hindbrain, and on top a series of bumps. These bumps, arranged as four pairs in a line, are known as *colliculi* (from the Latin for "little hills"). The first pair, on the forebrain, deals with olfaction (or sense of smell). The second pair is involved in reflexes dealing with vision and eye movements and the third pair processes sound. The second and third pair are found on the midbrain swelling. The fourth pair processes positional information, (called proprioception) and is found on the hindbrain swelling. This organization can be clearly seen in all animals along the evolutionary tree between fish and man but the relative sizes of the pairs of bumps vary, so that in some animals they are large enough to be called lobes, and in others, hemispheres. In early animals, information derived from smell became highly important for ensuring survival, as did the ability to process vision. Signals about body position and the resulting co-ordination were also important. As a result, the first and second pairs of bumps evolved to be larger, becoming lobes, with the fourth pair also expanding but to a lesser extent. As movement became more and more important, the fourth pair of bumps also began to expand into lobes. For mammals, the sense of smell and ability to move well are most important for survival and the first and fourth pairs have enlarged dramatically. In humans, this process of expansion of the lobes has gone to extremes. The first pair of lobes, originally dealing with smell, has enlarged gigantically, to become the cerebral hemispheres. In this expansion, they first grew forwards, over the forebrain, until they were confined by the front of the skull. They continued to expand, looping backwards over themselves until restricted by the back of the skull. They then had to loop forward over themselves again, this time at the sides, giving each hemisphere of our brain its familiar "boxing glove" shape and stretching the internal cavities (lateral ventricles) into a spiral.

In humans, the original olfactory function of the first pair of lobes is now confined to the inside and under-surface of the front portions of the hemispheres. This is where the sense of smell is located; it has important links with the emotion-controlling limbic system. The remainder of each hemisphere has taken over the functions of the vision and sound lobes, as well as containing areas for movement and sensation. The second and third pairs of bumps are now largely redundant, and are just small swellings on the back of the midbrain (the superior and inferior colliculi). They deal with visual- and hearing-based reflexes. The fourth pair of lobes, meanwhile, has developed and expanded further and is now about an eighth of the size of the rest of the brain. This is the *cerebellum*, or "little brain", and is essential for complex, learned, co-ordinated movements and their integration with information about balance, joint position and space. We therefore have a very large, sophisticated brain, specialized for processing sensation and turning it into action. Despite this, it is still not clear why it needs to be so large. Many successful animals have very small brains: for example, dinosaurs, with arguably the smallest brain size compared with body size the world has ever seen, were very successful.

clues from our cousins – apes and people

The first primates lived in Africa and this is where the first humans evolved. All primates share some features: forward-facing eyes, the ability to grip and leap, and nails instead of claws. In addition, humans have some unusual physical features. We are hairless, we have a truly opposable thumb (we can touch all our fingers with it), we have a downward pointing nose and we stand vertically. Socially we like to live in groups of about thirty and we follow a complex set of rules governing inter-personal relationships, which are learnt as we grow up. Primates spend a significant amount of time grooming each other. In humans this is done both physically and through the use of social rituals such as greetings. The physical features characteristic of humans are clues as to why we have large brains and our social abilities a consequence and cause. A side effect of living in groups is that we have developed the ability to imagine another being's viewpoint and to deceive. These social and mental skills are completely dependent on our brain power and are probably the evolutionary driving force behind the development of a large brain, as well as the result.

social winners and losers

Apes share a common social structure. Typically, a group of about thirty is led by a dominant, alpha, male, who has the mating rights over the group's females (but only if they accept him). He is supported by his deputies but usually opposed by rivals in the same group. Alliances are demonstrated by grooming, body-language and physical proximity. There is thus a constant battle for power, with the balance being held by females and a need to make the right choices about who to support. Making the right alliances at the right time can make the difference between successful fatherhood and celibacy. This requires the ability to understand complex social interactions, to understand the intentions of others and to control the amount that others can understand one's own intentions. It seems likely this has given us the ability to see the world from someone else's viewpoint (known as having a "theory of mind")

WATER BABIES – THE AQUATIC APE

We are alone among the great apes in actively enjoying being in water, and join with sea-dwelling mammals in having the ability to cry. We are born with a natural diving reflex, which, on the contact of the face with cold water, slows the heart rate and metabolism. In infants this is also combined with a reflex producing natural, anti-drowning, breath-holding and swimming movements. Is there a single model of what has happened in our evolutionary past which can explain all these features? One particularly intriguing theory is that of the "aquatic ape". This is based on the observation that many of the differences between ourselves and our ape cousins can be explained by an ancestral past spent in or near sea and fresh water. Wading through water requires a vertical posture, freeing up the arms, and some apes are observed to do this in the wild. Having free arms allows the development of hands that manipulate objects rather than simply being used for locomotion. Our body proportions are a 1:10 ratio, width to length; perfectly streamlined for water. Forward facing nostrils are prone to letting water in and, unusually for apes, ours face downwards. Large amounts of hair lead to waterlogging and again, unusually for apes, we are largely hairless. A diet high in fish provides the necessary essential dietary components required to build a large brain and it is generally accepted that our evolutionary ancestors must have had access to fish. Crying, in sea mammals, provides the vital function of removing excess salt from the body. Among apes, and among land mammals, we are unique in having this ability. This type of adaptation in our evolutionary past could explain many of our physical features.

There are other explanations for these features. The astronomers Hoyle and Wickramsinghe proposed that life originated in outer space, because there the chemicals necessary for the development of complex molecules are abundant. They proposed that comets should have an organic sooty core, an idea that was considered ridiculous because they were thought to be lumps of rock and ice. We now know they were right. Part of this theory also proposed that colds and flu epidemics sweep the world so rapidly because the viruses are arriving from space and that our downward facing nostrils are a way of reducing the entry of these viruses. This idea seems ridiculous . . .

and the ability to lie, as well as the need to communicate such concepts. These are the ingredients of consciousness. The other requirement is a brain of sufficient size. But how big is big enough?

the battle for bigness – brain size and brain power

An ant is capable of quite complex behaviour and yet its "brain" is barely the size of a pinhead. A mosquito can fly, home in on a target and extract blood, with just a dot of a ganglion. A rat is quite intelligent, but has a brain of only about two grams in weight – similar to that of a small part of the human brain, called the hypothalamus. In general, the larger the brain the greater the intelligence. This is not a strict relationship, however, and in fact it may be surface area which is more important, or brain weight as a proportion of body size. The convolutions on the surface of the brain are called *gyri* and the dips between them *sulci*. In humans the brain is not only very large compared with body size but is also highly wrinkled, giving it its "walnut" appearance. This means we not only have a relatively over-large brain, but also a surface area that is very large. In the cerebral cortex alone, an adult human has ten to twenty billion neurons and sixty trillion synapses. So where does all this extra brain power go? Is our behaviour really that much more complex than, say, a dog's or a pig's? The answer is, probably, yes. First, our interpersonal relationships are very complex. Second, we have a highly developed language ability. Third, we have a developed sense of time, both future and past. Fourth, we are capable of abstract thought. Fifth, we can place ourselves in the position of another person. Finally, we are capable of creating artistic works and solutions to problems. The combination of these is uniquely human, although it seems likely that some other primates and social mammals may be capable, to a lesser extent, of some of them. It seems that the critical components needed for brain power are numbers of neurons, numbers of interconnections and numbers of specialized circuits. In general, a bigger brain will provide these in bigger quantities, but this is only a general rule and, as always, there is an evolutionary trade-off between abilities and power on the one hand and maintenance and cost of upkeep on the other. Our brain uses up twenty per cent of the calories we take in, needs a long gestation (which is still not long enough, so we are helpless after birth for a long time) and needs a large head, making childbirth difficult and dangerous. This is a huge evolutionary burden, so it must be worth it.

intelligence – a unique evolutionary event?

We are often told that, as humans, our intelligence is what sets us apart from the other animals. Although this may be true at the moment, current thinking is that there have been several parallel branches of hominid evolution, only one of which gave rise to *Homo sapiens*. The evidence suggests that Neanderthal humans were highly intelligent, had language and tool making abilities, existed at the same time as our direct ancestors, but were not the ancestors of modern humans. This suggests that intelligence has arisen at least twice, once in modern humans and once in Neanderthals and we know there have been other early humans that also died out. We also know that dogs, for example, can understand hundreds of words, as can some horses and birds, and that some apes can communicate in sign language. It seems that intelligence is not such a unique event after all.

A brain is a vast collection of nerves, all interconnected in such a way that we can see and experience the world, think about it and act on what we conclude, but what are nerves and how does the brain work?

nerves – the body electric

. . . for no man lives in the external truth among salts and acids, but in the warm, phantasmagoric chamber of his brain, with the painted windows and the storied wall.

Robert Louis Stevenson (author, 1850–1894)

cell structure – soapy salt water is what we are

It comes as a surprise to many people to learn that we run on electricity. A human being is largely water with a few salts thrown in. We conduct electricity superbly, but not along wires. Nerves and muscles pervade every part of us and are the main routes along which current flows. In fact every cell, not just nerve and muscle cells, is electric.

All living things are built of cells. These are not just passive building blocks; they are busy active places. To understand the complexity of a cell, imagine an enormous underwater factory floating in the sea, vaguely round in shape. The factory walls are a gigantic soap bubble, perforated by doors and pumps. Although some water and small sea dwellers can leak in and out through the bubble wall, anything important or large has to come in through a pump or a door. The outer layer of the bubble is bristling with antennae and communications panels, sending and receiving signals. Inside, the factory is full of water, but, unlike the sea outside, this water is not salty. Any salt is pumped out as fast as it can leak in. Huge machines direct materials to different parts of the factory, messengers rush everywhere switching things on and off and gigantic conveyor belts move the factory's produce out to the appropriate door. Huge scaffolds connect and support everything and act as the local monorail. Just as this factory is a gigantic bubble floating in the sea, smaller buildings, also soap bubbles, float inside it and

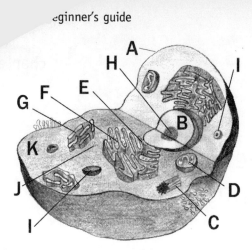

Figure 1 Diagram of a cell. A: cell membrane, B: nucleus, C: centrioles, D: mitochondria; E: smooth endoplasmic reticulum, F: rough endoplasmic reticulum, G: Golgi apparatus, H: chromatin (DNA), I: lysosome, J: free ribosome, K: cytoplasm.

all the bubbles are connected to each other through narrow soapy corridors. Each bubble has a specialized task, helping to keep the whole thing going. One type of bubble is a powerhouse, supplying energy, and there are usually many of these in each cell. Another is a large bubble, usually in the centre, where the product design team works. Their work is the reason the whole factory stays in business. They guide what needs to be done, and have all the plans needed to run the place, and to build a new factory if needed.

Now imagine that the factory has shrunk down to a fraction of a millimetre in size. That is a cell. The soap bubble wall is the cell membrane, made of a fatty substance with properties similar to those of washing-up liquid, so the soap bubble analogy is quite accurate. It can dissolve in fat or water. If it is in water, the fat-soluble parts are hidden and if it is in oil or fat, the water-soluble parts are hidden. Just as a soap bubble surface flows, so the cell membrane is quite fluid. You can imagine that a soap bubble might pop quite easily, so the membrane is strengthened with special proteins, continually replenished. Most cells also have an internal scaffolding, the cytoskeleton, to maintain their shape or help with movement. Although cell membranes are quite permeable to water and small molecules, they are full of pumps and pores which carefully control most of the substances passing in and out of the cell. The cell membrane is therefore described as semi-permeable. The antennae and communications panels on the outside are cell surface receptors. These are specially shaped molecules, each waiting for a specific partner molecule to attach itself. The partner molecule may be released by another cell or may be floating in the cell membrane. As soon as the partner molecule attaches, the receptor is activated. Like a key turning a lock, or a finger pressing a button, a cascade of events is set in motion. This is usually either the opening of a door allowing something specific into the cell, or switching on a message to be relayed within the cell. Messages are relayed to second messengers within the cell, which activate or de-activate the cell machinery. Living inside the cell are much smaller cells, which are bacteria called mitochondria. The mitochondria supply all the power for themselves and the main cell (the powerhouses of the factory analogy). This ancient partnership probably started many hundreds of millions of years ago, when a primitive cell tried unsuccessfully to engulf and digest a primitive bacterium. The cell survived and so did the bacterium, each living with the other. The cell was able to benefit from the power produced by the bacterium and the bacterium could benefit from the nutrients provided by the cell. Mitochondria are now an integral part of all animal cells, efficiently turning sugar and oxygen into power and neither the mitochondria nor the cell can survive without the other. The large central bubble-within-a-bubble is the nucleus, the design team's office. The nucleus contains the DNA blueprint of the organism. The relevant bits of DNA are unravelled and read to produce cell components, as well as being copied in their entirety when the cell divides.

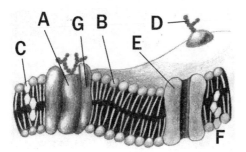

Figure 2 The cell membrane. A: protein, B: phospholipids, C: cholesterol, D: receptor, E: ion channel, F: cytoplasm, G: supporting proteins acting as a scaffold.

wired for sound, light, smell, anything – nerves communicating

Before explaining the nature of nerves, it is important to define what we mean by "nerve". To the general public, "nerves", "nervous" or "nervous breakdown" can be used to describe emotional states of anxiety or psychosis. A "nerve" is a part of the body that can lead to numbness or weakness if damaged. To a scientist, a nerve is a nerve cell or neuron; a single cell. To a surgeon or pathologist, a nerve is a yellow-white string-like bundle, made of many nerve cells. We will use "nerve cell" or "neuron" when we refer to the single cell and "nerve" when we refer to the much larger bundle of nerve cells.

In nature, there is a strong relationship between the structure of a thing and its function. Nerve cells are a particularly striking example. Because of the need for an extensive communication network, the usual round shape of the cell is deformed into a structure more like a sea urchin; its spines extended into long, delicate filaments. As the nerve cell grows, its filaments (called *dendrites*) seek out other nerve cells to contact and communicate with. Most nerve cells also have at least one extremely long, major, tube-like extension, known as an *axon*. In humans, an axon can be more than a metre long – like having a factory in Britain, about thirty metres across, with its own private tunnel stretching all the way to the USA. You can imagine that nerve cells need a specialized transport system, with its own type of trains and tunnelling materials, and indeed they do. In addition to the usual scaffolding and support system that all cells have, nerve cells have their own, smaller, scaffolding components: neurofilaments. These form the tracks for molecular motors to transport their cargo up and down the axon.

The more swollen, rounded, part of the nerve cell is the *cell body*. The cell bodies of nerve cells tend to group together and these groups form the grey matter of the brain and spinal cord. From the cell bodies,

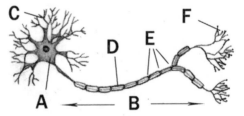

Figure 3 A nerve cell or neuron. A: cell body, B: axon, C: dendrites and dendritic branchlets, D: myelin, E: node of Ranvier, F: terminal boutons – these form synapses with other nerves or with muscles.

the axons shoot off to their own destinations, but they too are bundled together, to form a nerve. For example, the nerve cell that connects to the muscle in our little finger has its cell body in the grey matter of the spinal cord in the neck. Its axon extends down between the collarbone and ribs, under the arm, over the elbow and through the forearm to the little finger. The bundle of axons that go to the little finger muscles and skin form the nerve that causes tingling if we bang our funny bone. These axons are about a metre long. In the case of a blue whale, the axons may be tens of metres long. So, a nerve cell may have a very distorted shape, only a few thousandths of a millimetre across but many metres in length.

But the most amazing thing about nerve cells is their physical properties. A nerve cell can convert a signal of almost any kind into an electrical current. The commonest transformation is from chemical to electrical, but nerve cells can also change light, sound, temperature, pressure, stretch and even the earth's magnetic field into an electrical signal. The signal starts in the cell body and travels down the axon to be sent on to the next neuron in the chain, or it goes the other way, from axon to the body of the nerve cell. In some sensory nerves, there are two axons. In these neurons, the signal travels up the first axon, into the cell body and down the second axon until it reaches the far end, where it can be transmitted to other cells. You might think the easiest way for nerve cells to communicate would be by passing the electrical signal directly from one to the other, rather like an electrical circuit, but this is not what happens. When the signal reaches the far end, the nerve cell converts it into a chemical that diffuses out into the gap between it and its neighbour. Because it is a transmission of information from a neuron, this chemical is called a neurotransmitter. This region, where the neurotransmitter is released, where one cell ends and the next starts, is known as a *synapse*. When the chemical signal is detected by a nerve cell on the other side of the synapse, it is converted back into an electrical signal and the whole thing starts again. This mixture of electrical and chemical signals is called an electrochemical system and, combined with the physical arrangement of neurons and their connections, is how the nervous system works.

Nerves do not conduct electricity like a wire. Instead, they maintain a difference between salt concentrations inside and outside the cell membrane. To understand why this might make electricity, we need to understand what a salt is.

Elements can be grouped into two types: those that tend to lose electrons when combining with others (thus becoming positively charged) and those that tend to gain them (thus becoming negatively charged).

A salt is made of a pair of such opposite elements, in which the excess electrons of one have been "borrowed" by the other, forming a bond between them. In water, the elements dissociate, becoming positively and negatively charged "ions". In organisms, the most common positive ions are sodium (Na^+) and potassium (K^+). Sodium tends to attract electrons more strongly than potassium and, because the cell pumps sodium out but potassium in, a charge builds up across the cell membrane, with a difference of about 70 mV between the inside and the outside of the cell.

When a neurotransmitter signal is received from a neighbouring cell, the cell membrane becomes leaky to sodium, so a current of sodium ions begin to leak in faster than they can be pumped out. This makes the inside more positively charged. Because of the nature of the cell membrane, the more positively charged the inside is, the more leaky to sodium the cell membrane becomes. Gradually, more and more sodium leaks in and the inside becomes more and more positive. When a critical threshold is reached, specialized sodium gates open, letting huge amounts of sodium ions flood in. That region of cell membrane becomes strongly positive and depolarizes the neighbouring region, until it too opens the gates and depolarizes. The wave of depolarization spreads along the membrane, making the next part leaky, which depolarizes it, making the next part leaky and so on. In this way, the electrical current spreads down the cell membrane, never becoming weaker. This is known as an *action potential*. It is important to realize that there are three stages being described here. The first stage is the normal resting situation. The second is the initial leakiness of the cell membrane to sodium ions. The third is the catastrophic, massive leakiness to sodium ions that is the action potential, which only occurs when a critical threshold is reached.

The action potential method has two immediate advantages. First, a simple form of addition is possible. If a small signal is received, it might not be enough on its own to depolarize the cell membrane sufficiently to trigger an action potential, but many small signals together might. Second, because the signal never degrades, it can travel long distances without the need for repeater stations to boost the power. Compare this with the National Grid, in which large transformers are needed to maintain the electrical voltage because of the drop in power that happens in wires.

The main problem with the action potential method is that the speed of conduction of the electrical signal depends on the diameter of the nerve cell. Bigger is faster, but it is still painfully slow. This is why it can take a tenth of a second for you to feel it when you stub your toe.

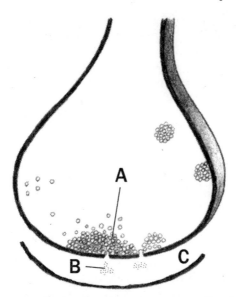

Figure 4 A synapse showing the release of neurotransmitter into the synaptic cleft. The neurotransmitter is stored in vesicles and released when an action potential allows calcium to enter the terminal bouton of the axon. The neurotransmitter then diffuses across the synapse to activate receptors on the far side. A: vesicle, B: neurotransmitter, C: synaptic cleft.

Higher animals have developed a solution to this problem. The axons of larger nerve cells are wrapped in insulation, giving the axon the look of a string of sausages. This insulation increases the speed at which the signal travels – it " hops" along the axon, rather than travelling in a continuous wave. This is called saltatory conduction, and it is when this goes wrong, for example, in multiple sclerosis, that it becomes clear how important it is.

When the signal reaches the far end of an axon, rather than the gates letting in sodium, they let in calcium, which activates the release of the chemical neurotransmitters into the synapse, ready to begin another action potential in the next cell.

neural nets – nature's internet

A few nerve cells connected to each other form a neural net. A slightly larger collection is called a ganglion and we usually refer to the largest ganglion or collection of ganglia in an animal as a brain. One of the

MUSCLE CELLS

Muscle cells conduct electricity in a similar way to nerves, but they differ in two crucial aspects. First, they are tightly connected to each other. In heart muscle, they are so tightly connected they actually merge together (called a syncytium). This allows the electrical depolarization to spread directly through the entire muscle. Second, the depolarization causes calcium to leak into a specialized membrane inside the muscle cell that contains proteins which can contract. Calcium activates the contraction process, so that an action potential causes a muscle cell to contract, which then spreads through the rest of the muscle. In heart muscle, the electrical pulse is set by a group of pacemaker cells that leak sodium at a preset rate, triggering an action potential about once a second that spreads to the rest of the heart: our heart beat.

simplest collections of nerve cells in nature is the neural net of *Hydra*, a small coral-like polyp that lives in the sea.

Because of the difficulties studying a neural net in a very small sea dwelling animal, scientists have made computer programs that have the same characteristics. Modelling simple neural nets with a computer shows that these structures automatically learn. One type of simple neural network is the three layer back propagation perceptron. This has an input layer, equivalent to the sense organs of an animal. These "nerve cells" are connected to the next (or hidden) layer which does the "learning". Finally the hidden layer connects to the output layer. A neural network is not like a computer because it cannot be programmed. Instead it is trained and tested repeatedly, until it has learnt. This is done by providing an input and "informing" the network whether is has produced the correct output. Based on this information, the strength of connections in the hidden layer is adjusted automatically by the network, because of the arrangement of learning rules programmed into the perceptron. This process is repeated until the training is over and the network can then be tested to see how well it has learnt. For example, the network might be taught to recognize letters. In the first step, a camera is shown a letter. The electrical signal from the camera feeds into the input layer of the neural network. The hidden layer processes this signal and in turn makes an output signal. The output might be connected to a simple panel of letters. If the correct letter is chosen, the connections in the hidden layer are strengthened; if the wrong letter is chosen, they are weakened.

In the same way as humans learn from other humans, the neural network needs a teacher to tell it when it is right and when it is wrong. It is important to remember that the teacher does not program the network, but only provides information as to whether the output was correct or not. Gradually, the network learns that a particular input corresponds to a particular desired output. In a sense, the input begins to have meaning. Seeing the letter A means it needs to choose the letter A. Such systems are excellent at recognizing patterns. Even a partial match will produce the correct output, much as we can recognize a face even if we only see a part of it.

Remarkably, if a part of a neural network is damaged, other parts will take over the function of the damaged region. This is rather like a damaged brain in which other regions take to speed recovery. It is also important to realize that although this learning process is taking place in a computer in the example above, it is identical to the process taking place in a physical collection of nerves arranged as a neural net. If sea polyps could be wired up to letter input and output devices (and had a "teacher" available), they would theoretically be able to learn in the same way.

It is not immediately obvious from knowledge of how a single nerve works that a collection of interconnected nerves would have the ability to learn or pattern recognize so effectively. A system like this, where complex behaviours arise from the relationship between simple entities, is said to have emergent properties. We will see what happens when this process is taken to extremes in the chapter on consciousness.

the brain as a computer

Because we do not understand the brain very well we are constantly tempted to use the latest technology as a model for trying to understand it. In my childhood we were always assured that the brain was a telephone switchboard. ("What else could it be?") I was amused to see that Sherrington, the great British neuroscientist, thought that the brain worked like a telegraph system. Freud often compared the brain to hydraulic and electro-magnetic systems. Leibniz compared it to a mill, and I am told some of the ancient Greeks thought the brain functions like a catapult. At present, obviously, the metaphor is the digital computer.

John R. Searle (*Minds, Brains and Science*)

CYBORG

In 1999, a group of researchers at Emory University and Georgia University, led by Bill Ditto, connected microelectrodes to leech neurons growing in a Petri dish, which were allowed to make their own connections. They found that this neural net could be trained to perform simple calculations. They dubbed this the "leech-ulator". In 2003, US researchers, led by Steve Potter, connected a dish of neurons, via the Internet, to a robot arm in the laboratory of researchers in Australia. The idea of the project, called MEART (multi-electrode array art) was to explore the creative process. This "brain" and "body" was designed to draw artistically. The neurons received as input a video image of a visitor to an art gallery and another of the page, so they could "see" what they were drawing. Although the images were far from accurate, they became more organized and less chaotic with time. More recently, the laboratory of Miguel Nicolelis at Duke University has developed a device allowing a person with a spinal cord injury to control a computer cursor. The device is implanted over the motor cortex and the electrodes interface with layer-three neurons in the brain. Information is collected from the device by a magnetic reader held over the skull so there is no need for any part of the skin to be broken. Initial experiments in a man with a spinal cord injury have proven successful, although the work has been controversial because initial studies were performed on primates. In 2004, researchers at MIT and others in New York state, developed systems using brain wave patterns picked up from special caps containing electrodes. With these, the wearer was able to control a computer game or move a computer cursor, purely by thought control.

It is possible that systems like these will lead to the eventual integration of the human nervous system with electronic systems. This would make humans of the future true cybernetic organisms.

The human brain is organized in cell layers up to six deep, each with thousands of different inputs and sending its output to huge numbers of other neurons connected in a complicated three dimensional pattern. The different networks are themselves connected to each other in complicated patterns and the outputs are affected by global changes in the general chemistry of the brain. It is a neural network fantastically organized and far more complex than any we have yet conceived.

If collections of nerves can learn to produce a particular output for a particular input, does this mean the brain is nothing more than a computer? The answer to this depends on what we mean by "computer". Most people use this term to describe a machine that can be given a list of instructions (a program) which gives it a new ability, for example displaying a picture on a screen or behaving like a calculator. This is quite different from a neural net. For a computer, there is no training involved, only a program. Given the same input a million times, a computer will normally produce the same output a million times, unless it has been programmed not to. Give a neural network the same input a million times and you may never receive exactly the same output. Computers are not very good at recognizing patterns and have to use statistical techniques to mimic this. Neural networks are superb at recognizing, but they cannot be programmed, only trained. Even simple neural networks are not computers and highly complex nets like the human brain are certainly different, although it can be helpful to think of the brain as composed of connected thinking "mini-computers", as we shall see later. The biggest difference is probably philosophical. A computer does not need to interact with its environment to "know" what to do. It simply follows a list of instructions. In this sense its actions have no meaning and it has no understanding. On the other hand, a neural net has to interact with its environment to learn what to do, which argues that a particular response to what it sees has a meaning.

We have seen how a single nerve cell works and that a small network of nerves has emergent properties allowing it to learn. In organisms a little more advanced than *Hydra*, nerve cell bodies collect into groups called ganglia. The largest ganglion is usually at the head end of the organism. This is the smallest unit that can be thought of as a brain and it is a glorified ganglion that we have in our heads. Although everyone is different, our brains all seem to follow the same general plan, even though they contain many billions of nerves, each with many thousands of connections. How can all this arise from a single cell?

making a brain and mind from one cell

the development of the brain – growing a nervous system

A hen is only an egg's way of making another egg.
Samuel Butler (composer and satirical author, 1835–1902)

from egg to embryo

Our nervous system's development reflects a little of our evolutionary past. To understand it requires the ability to imagine objects morphing from one form to another, rather like the evil T-1000 terminator played by Robert Patrick in the film *Terminator 2*. If we apply some virtual time-lapse photography, we can describe the development of a person from day one as a fertilized egg to day 266 as a newborn, rather like speeded-up film of flowers opening or clouds racing across the sky.

Once we come into existence, an incredible controlled explosion of the highly compressed information held by the fertilized egg begins; a cascade of events leading from a single cell to an entire living, functioning human being. This is the true miracle of life. The initial event is the very first cell division, from a single cell to two cells, which takes place about thirty hours after conception. Cell division continues for the rest of our lives in various forms, but to start with it consists of a simple doubling of cells at every division, which lasts for about three days until there is a solid sphere of cells. Over the next few days, spaces appear in the sphere, which gradually coalesce to form two large spaces. While this has been going on, the developing embryo has been travelling down the Fallopian tubes towards the uterus. Once there, the sphere burrows into the wall of the uterus and begins the task of becoming a person.

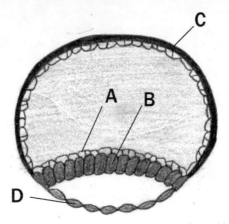

Figure 5 The early embryo, or blastocyst. A: top surface which will eventually form most of the adult, B: lower surface, C: will become the membranes enveloping and protecting the fetus, D: will form the placenta.

The early embryo resembles two hollow hemispheres stuck together. The two flat adjacent surfaces will eventually form the body, the hemisphere that burrowed into the uterine wall will become the placenta and the other will become the membranes that encase and protect the developing fetus.

the primitive streak – making heads or tails of things

At this stage, the embryo is a two-layered disc in the centre of a sphere. However, from now on we will imagine the disc alone and ignore the sphere, because the disc is what will become a person.

At this stage, the embryo has a definite top and bottom surface, the top being the layer nearest the membrane hemisphere and the bottom being the layer nearest the placenta. At around day fifteen, a streak of new cells appears in the top layer. This is the *primitive streak*. It starts out as a darkening at one edge of the disc (the *primitive node*) and spreads to become a straight line through the centre to the other edge. As the darkening spreads, it forms a groove and at the far end it finishes in a small depression (the *primitive pit*). This will become the head end; the other, where it started, the tail end. The primitive streak is now the midline, defining left and right to either side. Over the next day, a group of cells on either side of the primitive streak streams down through the groove and completely replaces the bottom layer of the disc. This bottom layer of cells, the *endoderm*, gives rise to the lining and major organs of the gut (the liver and pancreas). Another group of cells streams down and positions itself in between the two layers of the disc, to form a third layer. This, the *mesodem*, gives rise to muscle, the skeleton, kidneys, bladder, genital system, some of the coverings of the internal organs and the deeper layers of the skin. The topmost layer, the ectoderm, gives rise to the rest of the skin and the nervous system. So, by day sixteen, the embryo looks like a three-layered disc, with a groove running the length of the topmost surface, one end of the groove being slightly deeper than the other.

the notochord – the defining moment for the nervous system

While this is happening (day sixteen), the group of cells around the primitive pit, at the head end of the primitive streak, also migrates down into the middle layer. These cells travel even further and join with the endoderm of the bottom layer. There is therefore, at this stage, a hollow tube of cells from the top surface through to the bottom surface of the embryo at the head end. The very bottom end of this tube becomes the mouth. The rest solidifies and migrates down and along until it only occupies the middle layer and forms a solid rod along the length of the embryo down the midline. This is the *notochord*. It is the precursor to a backbone and is found in all vertebrates (and in many more primitive animals, the most primitive being the lancet). In adult mammals the remnants of the notochord are found in the discs between vertebrae. The presence of a notochord distinguishes animals that have a nervous system organized like ours from those that do not (such as jellyfish or insects).

The mesoderm cells that migrated down from the outer region around the primitive streak have in the meantime moved sideways and begun to form circular structures that will become the main groups of muscle. These structures are regular and form in segments (in the same way an earthworm is segmented). The segments can be numbered and, for example, the first seven will form components of the head. At this stage, the embryo looks like a sandwich with a filling made of a single sausage with lumps of meat on either side. The top layer of bread is the ectoderm, the bottom the endoderm, the lumps of meat the mesoderm and the sausage the notochord.

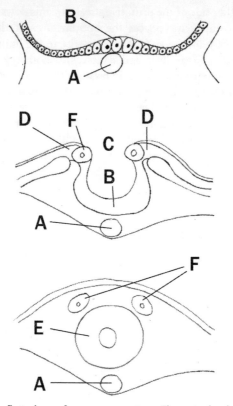

Figure 6 The first signs of a nervous system. The notochord develops and causes the ectoderm above to become neural tissue, which in turn sinks down to become a tube of nervous tissue – the future spinal cord and brain. A: notochord, B: neural plate, C: neural groove, D: neural crest, E: neural tube, F: future muscle.

the neural tube – the first steps to thought

At around day eighteen, by releasing chemical signals, the notochord initiates the formation of the nervous system, by inducing cells in the ectoderm to mature a little and become precursor nerve cells. It also signals them to start replicating. Thus begins the production of the one hundred billion nerve cells of the adult brain, at a rate of 250,000 neurons per minute. The first step in this process is the formation of the *neural plate*. In response to a notochord protein called "Sonic Hedgehog", this region of ectoderm gradually sinks down to form a

SPINA BIFIDA

Failure of the neural tube to zip up leads to conditions known as neural tube defects. The best-known of these is spina bifida, a condition in which the section at the lower end does not fuse. Less well-known is anencephaly (from the Latin for "no head") in which the top end fails to zip up. Neural tube defects are more likely if the mother had a deficiency of folic acid before and during pregnancy and, as a result, pregnant women are advised to take folic acid supplements: in the USA, it is added to flour.

groove (the *neural groove*) and at the same time the regions either side rise up to form a crest (the *neural crest*). The cells nearest the notochord, which receive the biggest dose of "Sonic Hedgehog" will become motor nerve cells. Meanwhile, on the top surface, the crests come closer together, while the groove sinks further, to become a tube, over which neural crest cells "zip up". The zipping-up starts at the fourth segment and spreads down and up until only two openings are left, one at each end of the embryo. On day twenty-four, at the head end, closure begins and completes at the level of the future forebrain. At the tail end, the opening begins to seal up around day twenty-six and completes its closure at the level of the second sacral vertebra, below the small of the back.

By day twenty-eight, the neural tube is completely closed. The hollow centre will become the spinal canal and ventricles of the brain. As the neural crest sealed, some of the excess cells pinched off to lie on either side. These migrate through the embryo to form the peripheral and autonomic nervous systems, some hormone releasing glands, such as the thyroid and adrenals, and the pigment producing cells of the skin (melanocytes).

segmentation of the neural tube – sculpting a brain

From day thirty to day sixty the brain begins to develop its more familiar shape. The neural tube first develops three swellings at the head end. These form the *forebrain, midbrain* and *hindbrain*. The forebrain segments to form the *telencephalon* (end brain), which will become the cerebral hemispheres, including the cortex and deeper brain structures such as the basal ganglia, and the *diencephalon* (between brain), which

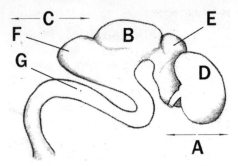

Figure 7 The embryonic brain. A: forebrain, B: midbrain, C: hindbrain, D: telencephalon, E: diencephalon, F: metencephalon, G: myelencephalon (medulla or hindbrain).

will become the thalamus and also give rise to the eyes. The midbrain remains unsegmented. The hindbrain segments into the *metencephalon* (behind brain), which will give rise to the *pons*, the *cerebellum* (little brain) and the *myelencephalon* (*medulla* or spinal brain). The medulla is the lowest part of the brainstem and is, in some ways, a continuation of the spinal cord into the skull. As the swellings develop they become more obviously paired and symmetrical structures, lying either side of the midline.

Rapid growth now folds the developing brain in three. The *cervical fold* bends the brain forwards between the spinal cord and the medulla. At the other end, the *midbrain fold* bends the brain forwards between the forebrain and midbrain. In between, the *pontine fold* bends the brain backwards, so that the midbrain is forced upwards. This also creates a wide space, by squashing out the central part of the neural tube, the fourth ventricle. The stretched layer of cells in its roof forms a thin membrane. The pontine fold segments the hindbrain into the metencephalon and myelencephalon. The brain at this stage is a little like a zigzag made of swellings in a tube, forming an M shape.

maturing the brain

The brain has now begun its journey into maturity. The task that lies before it is to grow in size as well as in numbers of connections. To do this, it will follow a set of genetic instructions which will be modified by experience. At the same time, the experiences it will have are determined by the set of connections, so there is a causal loop – brain wiring modifies experience and experience modifies brain wiring. For

NATURE AND NURTURE

The nervous system develops gradually. There does not seem to be an obvious time when it is switched on; a time that divides a non-thinking from a thinking state, or a time that separates a brain-silent from a brain-active state. The heart too forms gradually but does start spontaneously, to beat in a co-ordinated manner at around day twenty-two of gestation and therefore there is a moment at which it can be said to start working: there is a "first heartbeat". Is there a similar process for the brain? Nerve cells do "beat" – there is a co-ordinated signal that leads to rhythmical waves of activity, if no external signal is received. With the eyes closed, the nerve cells in the back part of the brain, the occipital lobe, fire at about 8–11 signals per second. This is the *alpha rhythm*. Opening the eyes destroys it, presumably because the cells are now processing visual information. Interestingly, the alpha rhythm is also destroyed during dreaming sleep, suggesting that for the occipital cortex the signal is real, something that is in keeping with our personal experiences. Deeper parts of the brain have a slower natural rhythm, the *delta rhythm*. In evolutionary terms, these deeper structures are more primitive and older. So, nerve cells fire with a rhythm equivalent to a heartbeat. This rhythm must start at some point in development but because measuring it requires the attachment of electrodes to the skull, we do not know when or how it begins. We also do not know if this signals the start of a functioning nervous system, although it seems likely.

example, someone born with a squint has eyes that point in different directions. As a result, it is impossible for the central vision of each eye to be fixed on the same object. To prevent double vision, the developing brain suppresses one of the images. Eventually, the suppressed eye becomes functionally blind, because the brain circuits to interpret the signal do not exist. Even if surgery is performed, there are no connections to "read" the signals coming from the eye. The brain wiring has been altered from its genetic program by the events of life (although this has happened in a genetically programmed way). Nowadays, to correct the problem, we cover the eyes alternately, until the child is old enough to have corrective surgery. This compels the brain to accept the images from both eyes and forces the wiring to grow correctly. Even so, because the wiring for three-dimensional vision requires simultaneous input from both eyes, depth perception will never be quite the same as

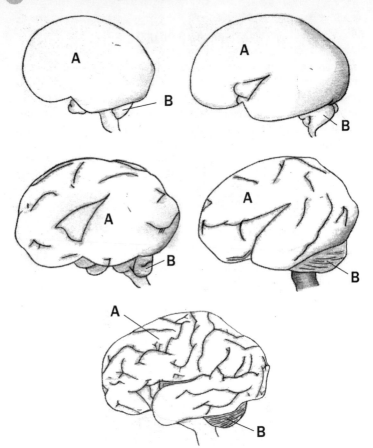

Figure 8 The maturing of the brain from 4–9 months gestation. As the developing infant grows, the cerebral hemispheres enlarge and overflow, covering the rest of the brain. The cerebellum also enlarges, protruding at the back. The temporal lobe becomes more obvious and the fissures, sulci and gyri of the adult brain develop. A: cerebral hemisphere, B: cerebellum.

if both eyes had worked together from the outset. This immediately begs the question that, if a lack of sensory input leads to a change in brain wiring, what happens if there is extra information? For example, if a child were connected to an infra-red camera through wiring directly interfacing with the developing brain, would the brain circuitry adapt and become able to interpret the images as if the camera were simply another sense organ? This is a fundamental question because the brain seems to have specialized areas able to process particular information. Individuals born with extra digits are able to use the extra digit normally. This means that the parts of the brain dealing with digits do not "know" how many to expect. The visual areas cannot "know" whether information will be in colour or not, or if there will be information from both eyes. Would they therefore care if information came from an infra-red camera? Development of different brain areas is probably loosely programmed and then modified by what the nerves find as they connect between themselves, with the motor organs and, via the senses, with the outside world. This ability to be moulded is known as *plasticity* and is still present, though to a lesser extent, in the adult brain. This is probably because the reconnections required by plasticity are easier in a developing brain than an established one.

As we will see in the chapter on the development of reasoning and behaviour, this idea of specialized areas that must respond in a particular way, compared with the idea of a completely flexible system, is central to our abilities. Before we get there, though, it is helpful to have an overview of how the brain is arranged and what each part does.

chapter five

the anatomy of the brain – understanding the grand plan

The brain is a wonderful organ. It starts working the moment you get up in the morning, and does not stop until you get into the office.

Robert Frost (poet, 1874–1963)

My brain: It's my second favourite organ.

Woody Allen (*Sleeper*)

No brain is stronger than its weakest think.

Thomas L. Masson (author, 1886–1934)

finding our way through the white and the grey

If you examine a man [having] a gaping wound in his head, reaching the bone, smashing his skull and breaking open [the viscera] of his skull, you should feel [palpate] his wound. You find that smash which is in his skull [like] the corrugations which appear on [molten] copper in the crucible, and something therein throbs and flutters under your fingers like the weak place in the crown of the head of a child when it has not become whole.

Anonymous (*c.*1550 BCE), in *Ebers papyrus* (cited in J. F. Nunn, *Ancient Egyptian Medicine*)

The adult human brain weighs, on average, about 1.35 kilograms (about three pounds). To understand the broad functions of the various parts of the brain it is helpful to divide it up into manageable chunks, starting with the largest part: the cerebrum. The cerebrum is a highly folded

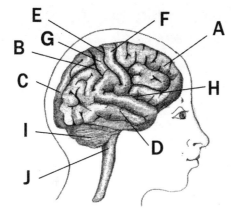

Figure 9 The lobes of the brain. A: frontal lobe, B: parietal lobe, C: occipital lobe, D: temporal lobe, E: central sulcus, F: motor strip, G: sensory strip, H: Sylvian fissure, I: cerebellum, J: brainstem and spinal cord.

structure in two hemispheres, which looks a little like a walnut. The folds are termed *gyri* and the grooves in between them *sulci*. The two halves are joined in the centre by a band of nerve fibres, the *corpus callosum*, allowing communication.

If we look at the hemispheres in cross-section, the central portion appears lighter and is referred to as "white matter". It appears white because of the presence of myelin, the insulator which allows nerves to conduct signals more quickly. (In diseases like multiple sclerosis, the myelin becomes damaged, seen as multiple patches of sclerosis or "scarring" throughout the brain.) Covering the white matter, like the rind of an orange, is the darker "grey matter" or *cortex*.

Because the hemispheres of the cerebrum are symmetrical, we will from now on only refer to one. The outer part, or cortex, can be divided into four regions, the *frontal, parietal, occipital* and *temporal* lobes. The temporal lobe is separated from the rest by a groove called the Sylvian fissure.

Deeper within the brain, towards its base, are the *basal ganglia*, regions concerned with the regulation of movement. Co-ordination of movements is controlled by the "little brain" or *cerebellum*, at the back end of the brain. Finally, the brainstem, a "stalk" arising from the base of the brain, is the exit and entry point for multiple nerve fibre pathways supplying the limbs and internal organs. It consists of three portions, the *midbrain, pons* and *medulla*. The medulla connects the brainstem to the spinal cord.

the cerebral cortex – using your grey matter

He who joyfully marches to music in rank and file has already earned my contempt. He has been given a large brain by mistake, since for him the spinal cord would fully suffice.

Albert Einstein (physicist, 1879–1955)

I not only use all the brains I have, but all that I can borrow.

Woodrow Wilson (US President, 1913–1921)

An idle brain is the devil's workshop.

English proverb

"Cortex" comes from the Latin for "rind", which demonstrates the rather dismissive view of this part of the brain in early teaching. Early ideas about brain function were somewhat wide of the mark and for centuries almost all investigators ignored the cerebral cortex. To be fair, the Greek anatomist Erasistratus had noted that human brains had more folds than those of other animals. Galen, however, scorned to attribute any significance to this and did not regard the cortex as particularly important. This view persisted until the eighteenth century. Thomas Willis (English physician and anatomist, 1621–1675; see also chapter 6) was a notable exception. Credited with founding the Royal Society and recognized as the author of the first monograph on the brain, Willis was the first to describe the circulation of the brain and one of the first to attribute memory and voluntary movement functions to the cortex. However, the dominant view of the time was that the cortex was simply a protective covering and the fact that directly injuring or irritating it does not cause pain was used as an argument against it having any important functions.

We now have a very different view. Our cortex is a defining part of our individuality and, in evolutionary terms, this part of the brain is the most developed. The adjuration to "use your grey matter" reflects its role as the seat of intelligence. Like all organs in the body, it has a remarkable functional reserve within it, which allows for brain plasticity, a process in which damaged parts of the brain can be substituted for by other regions.

motor and sensory cortex

The frontal and parietal lobes are separated by a large fissure, the *central sulcus*. Either side of this lie the motor and sensory cortex; the motor in the frontal lobe and the sensory in the parietal. (This split, between

USE YOUR BRAIN!

It's a popular myth, without any scientific basis, that we use only ten per cent of our brains. Given that other organs in the body have at least fifty per cent of their function "spare" (for example, one of each pair of the lungs and kidneys can be removed without appreciable effect), it is not unreasonable to think that the brain is the same. Certainly, for some brain diseases, many nerves have to degenerate before the loss is noticed by the patient. Furthermore, there are certain congenital conditions in which a large part of the space for the brain is taken up by enlarged ventricles, leaving only a thin rim of brain around the outside, yet this is not always associated with obvious problems in learning or development although formal testing would reveal detectable problems. Damage ninety per cent of someone's brain and they will certainly die. Even a very small stroke can be devastating. Given that the brain uses a fifth of the energy we take in, there will have been great evolutionary pressure to make it as efficient as possible. This means it is unlikely there is much redundancy in the brain beyond what is needed to allow us continuously to learn and remember. There is also some evidence that we should "use it or lose it"; people who use their brains more often, for example by doing crosswords regularly, are less likely to develop certain degenerative conditions like Alzheimer's. So if we only used ten per cent of our brains, they would probably be much smaller.

sensory inputs being dealt with in the back half of the brain and motor outputs in the front, is repeated for virtually every system.) Broadly speaking, incoming sensory information (for example, temperature, pain, touch) terminates in the sensory cortex and outgoing nerve impulses, concerned with the generation of movement, originate in the motor cortex just in front of it.

As the experimenters of the nineteenth century discovered, the remarkable thing about both the motor and sensory cortex is that information is laid out in a specific topographical map. For example, at the very top is the area for the leg, then, coming further down, those for the arm and face. Each part of the body has its own specific representation in the cortex but the proportion of nerve cells given to each part is not equal. This is evident if we consider the exquisite sensitivity and dexterity of the fingertips. If we draw a picture of the human body,

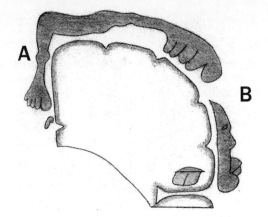

Figure 10 A map of the left motor or sensory cortex, with the section taken facing the individual. The different body parts are represented in predictable defined regions. A: midline, B: side.

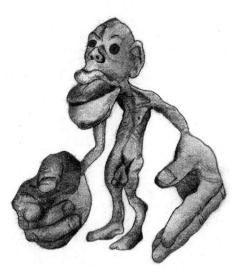

Figure 11 The homunculus, showing the relative amounts of cerebral cortex devoted to sensation for the body. This is why a small piece of grit feels enormous on the tongue, but when seen, is remarkably small.

weighting the size of each part of the body according to the amount of motor or sensory cortex given over to its function, the result is an odd-looking "homunculus", with enormous hands and lips.

PHANTOM LIMBS

The existence of the homunculus may account for the phenomenon of *phantom limb*; the perception that an amputated limb is still present. Phantom limbs are frequently very painful, or are perceived to be held in a distorted position. Even though they can be felt, they cannot be moved. For limbs that feel as if they are held in a tightly curled or stretched position, this can be quite distressing.

frontal lobes – suppressing the undesirable?

Some scientists are moving away from the simplistic view that separate functions can be very clearly assigned to different lobes of the brain. However, some broad categories of function can be ascribed to each of these regions. The large frontal lobes, for example, are involved in personality, as illustrated by the history of Phineus P. Gage (1823–1860). Gage was preparing a railway bed near Cavendish, Vermont, when, on 13 September 1848, having packed too much gunpowder into the charge, an explosion sent a 3 foot 8 inch long, 1.25 inch circumference, $13^{1}/_{2}$ lb metal tamping rod through his left cheek bone and out of the top of his skull. Landing some thirty yards away, it unfortunately took or destroyed most of Gage's left frontal lobe with it.

An article in the *Boston Post* reported: "the most singular circumstance connected with this melancholy affair is that he was alive at two o'clock this afternoon, and in full possession of his reason and free from pain". It is said that Gage did not even lose consciousness and was back at home several weeks later, apparently unimpaired. However, a dramatic change in personality had occurred. Before the accident, Gage was described as "dependable, industrious, and well liked", but when he recovered he was "restless, loud, profane, and impulsive". His doctor went on to describe him as "manifesting little deference to his fellows, impatient of restraint or advice when it conflicts with his desires, at times pertinaciously obstinate, yet capricious and vacillating, devising many plans of future operations, which are no sooner arranged than they are abandoned". His friends described him as "no longer Gage" and his former employers refused to reinstate him.

This story has now passed into the legends of neuroscience and Cavendish has a commemorative plaque. As well as being the first known example of self-administered brain surgery, it illustrates the

EPILEPSY

Groups of nerves normally show two patterns of behaviour: rhythmic waves of synchronized activity (usually a resting state) and chaotic-looking unsynchronized activity (an active state). A third state can occur, in which synchronized, rhythmic activity becomes overwhelming and includes nerve groups that would not normally be synchronized. In this state, the amplitude of the synchronized waves is much larger than normal. This is the electrical basis of epilepsy. If the electrical disturbance involves the brainstem, the person will become unconscious. If it includes the part of the brain dealing with movement, the arms or legs will convulse in time with the electrical waves, as the motor cortex discharges. An unconscious person convulsing is what most people recognise as epilepsy, but the electrical disturbance can remain confined to one portion of the brain and consciousness is not necessarily lost. In this case, the person will experience things based on the part of the brain affected. Reports of these experiences can be compared with brain scans and used to work out which part of the brain does what.

frontal lobes as major players in determining aspects of personality, normal social behaviour and inhibition. Consequently, diseases affecting the frontal lobes can manifest as loss of inhibition and inappropriate behaviour. Neurologists quickly learn to identify someone with this pattern of problem. The changes may be quite subtle; just a little too much laughing or joking, rudeness or impatience and almost child-like changes between the two. These changes are the result of underactivity of the frontal lobes, due to damage, but similar patterns occur if the frontal lobes are over-active, for example in people with frontal lobe epilepsy, who, during a seizure, may exhibit bizarre, uninhibited, behaviour such as inappropriate laughing or crying.

the temporal lobes – thanks for the memory

Much of our knowledge about the normal function of the temporal lobes has come from the study of people with epilepsy originating in this part of the brain. People with epilepsy may experience an *aura*, an awareness that can herald an impending seizure. In temporal lobe epilepsy (TLE), this occurs in the majority and may take many forms.

For example, some report a sudden feeling of *déjà vu* (although this can also be a normal phenomenon) or the opposite feeling of complete unfamiliarity or *jamais vu*. Some people experience intense fear, or become aware of an unpleasant smell or taste.

Because the brain itself has no pain receptors, brain surgery can be carried out, under a local anaesthetic, on an awake patient. Dr. Wilder Penfield (1891–1976), an American neurosurgeon, used an electrode to stimulate various parts of the temporal lobes in the brains of people with TLE. It caused them to "hear" sounds in the form of simple tones or even music, or altered the tone of sounds already played to them. A young girl undergoing stimulation of this brain area said that she was aware of something coming towards her. In another position, holding a bar over her head as she lay down on the table, she experienced the feeling that the bar was moving away, despite holding on to it. At another point of stimulation she reported voices shouting at her to do something wrong and experienced a feeling of impending doom. Other people recalled memories, which occurred one at a time, proceeded forwards and, like a short film, could be stopped and replayed by altering the stimulus. The person never lost touch with reality and was aware that these memories were playing alongside the present, as a result of stimulation rather then spontaneous recall.

As these experiments suggest, the part of the brain responsible for the processing of information from the ear, the auditory cortex, is within the temporal lobes. In the same way as in the motor and sensory cortex, it is thought to be organized topographically, shaped by exposure to a variety of sounds during early development. The temporal lobes also seem to be particularly critical for memory function. On the innermost surface is a specialized structure, the *hippocampus*. (This name comes from the Greek word for "sea horse", because of its shape in cross-section). The hippocampus is involved in the processing and packaging of memories for storage elsewhere. Its highly specialized nerve cells are particularly able to alter both their responsiveness to stimulation and their connections with other nerves. People with damage to the hippocampus have an impairment of memory, called *anterograde amnesia*, in which they cannot form new memories, but do have recollection of the distant past and can learn new skills with repetition, suggesting that the hippocampus is, more specifically, involved in encoding the "context" to a given memory. The extreme degeneration of the hippocampus has tragic results. Sufferers of Alzheimer's disease, which particularly affects the hippocampus of the temporal lobe, gradually become increasingly forgetful of routine tasks, frequently get lost and eventually do not recognize even close family.

Finally, what about religion? There is good evidence for a theory, put forward by the neuroscientist V. S. Ramachandran and others, that the temporal lobes are the source of religiosity and that religious beliefs are simply a product of our "wiring". Certainly, some patients with TLE do report hallucinations with a religious component and, as we shall see later, there is other evidence for this idea.

the occipital lobes – seeing the whole picture

The occipital lobes are the primary region for the processing of the vast amount of information arriving from the eyes. Humans are extremely reliant on vision, and a large amount of cerebral "power" is dedicated to this sense (see chapter 13).

the parietal lobes – making sense of sensation

The vast amount of sensory information from the other modalities of touch, hearing, smell and taste must be interpreted, to allow co-ordination with motor output, and this is one of the major functions of the parietal lobes. Damage to the parietal lobes leads to a total or partial loss of the ability to perform a purposeful series of movements, for example waving someone goodbye or combing one's hair, or of the ability to manipulate objects, even though there is no weakness or sensory problem.

The illumination of the more obtuse functions of the parietal lobes owes much to neuropsychology. The parietal lobes remain mysterious and ill-understood; much of our limited understanding has come from the study of patients with a variety of diseases of these areas. One of the most famous contemporary "celebrity" neuroscientists – the American neurologist Oliver Sacks – has documented many amazing stories in his fascinating book, *The Man Who Mistook His Wife for a Hat*, in which the man of the title lost the ability to recognize and differentiate accurately between certain objects, including his wife. Other stories describe people with "alien" limbs, beyond their control. Although not all the cases were exclusively caused by parietal lobe damage, this is undoubtedly the centre of sensory information processing.

lateralization – split personalities?

The right half of the brain controls the left half of the body. This means that only left handed people are in their right mind.

Anonymous

A curious aspect of the wiring of the brain that fundamentally affects its control is that nerve fibres to and from the major lobes cross the midline in the medulla or spinal cord, so that the left cortex controls the right side of the face and body and vice versa. This is why someone who has had a stroke in a blood vessel supplying the left side of the brain has weakness and numbness over the right side of the face and body.

It is perhaps not surprising to discover that the left and right halves of the brain are not equal. This is lateralization. Despite it, some people who have lost half of their brain, through injury or surgery, can function normally in many ways, but only if the damage occurs early in life, so that the remainder of the brain can take over the functions of the damaged part. Neurologists refer to the "dominant" and "non-dominant" sides of the brain: "dominance" is predominantly derived from the location of language functions. Almost all right-handed people are dominant for language function on the left side of their brain (which controls the right side of the body). Surprisingly, most left-handed people are also dominant for language on the left side – only in about twenty per cent is the right side dominant. This can become important in diseases such as stroke. A stroke affecting the left side of the brain is much more likely to affect language than one on the right-hand side. The loss or impairment of language after a stroke is, therefore, most usually in combination with right-sided body weakness. If there is left body weakness instead (that is, right-sided brain damage) with language impairment, then the patient is almost certainly in that group of left-handers with a "reversed brain".

When surgical removal of parts of the brain is being considered, for example because of a tumour or drug-resistant epilepsy, it is easy to see how vital it is to be sure that language regions are not likely to be damaged. If babies or children sustain damage to or have surgery involving these areas of the brain, there seems to be a capacity for the brain to "re-wire" itself and "move" these centres to undamaged parts of the brain, even to the opposite side, restoring function and near-normal development. The optimal time for this plasticity seems to be during the first two years of life; unfortunately, it declines with age. The understanding of how plasticity operates and the ability to harness its power to treat stroke patients or victims of head injury or degenerative diseases constitute the holy grail of modern neuroscience.

Although the left hemisphere is most often dominant for language, the non-dominant side may be involved in appreciating the emotional tone and significance of the spoken word and in the construction of an appropriate verbal response. For memory, left temporal damage tends to result in impaired memory for verbal material, whereas right-sided damage impairs recall of non-verbal material, such as music or drawing. Interestingly, this part of the brain retains a degree of plasticity into adult life, far more than other regions, so that functions of the damaged temporal lobe are often taken up, to some extent, by the opposite side.

Not only language and memory are lateralized. The parietal lobes are responsible for many of the more abstract functions, such as spatial processing and face recognition, which are also lateralized to dominant and non-dominant sides of the brain.

the split brain – two brains in one skull

Finally, as if to prove hemispheric lateralization, some bizarre and fascinating syndromes can arise from damage to the connection between the two hemispheres, the *corpus callosum*.

The splitting of the corpus callosum has sometimes been performed deliberately, in cases of severe epilepsy, to prevent the spread of abnormal nerve discharges from one half of the brain to the other. However, problems can arise. For example, if the two halves of the visual association cortex (in the occipital lobes) cannot communicate with each other, then the words of a book in one half of the visual field would be fine, but the words in the other half might seem meaningless. Or a person with their eyes closed might recognize an object (for example a key or a coin) placed in their hand but if the link to the other side of the brain was not working then they might be unable to name it.

Absence or abnormal development of the corpus callosum can occur naturally. Once again, babies seem to be far more resilient to this kind of abnormality and, although there can be learning difficulties, they can also develop entirely normally.

In the nineteenth century the eminent French neurologist Joseph Jules Déjèrine (1849–1917) had a famous case. A well-educated banker awoke to find he had lost the ability to read words, despite seemingly reasonable vision. Déjèrine found that the patient could still spell even the most complicated words, despite being unable to read even the simplest ones. He could write fluently but was unable to read what he had just written. Bizarrely, if he was able to trace the letters with his fingers, all would become clear. Although he could no longer read music, he was able to learn new pieces by listening to his wife singing the parts. His ability to read numbers was unaffected. A post mortem revealed he had damage in the corpus callosum – effectively disconnecting the two halves of his brain – but because the number-processing area lies within the temporal lobes, in a region where the right and left sides are connected via an alternative pathway, the *anterior commissure*, this remarkable pattern of symptoms arose.

the limbic system – smell the fear!

Along the lower edge at the tip of each of the temporal lobes is the *amygdala* (from the Greek word for "almond"). This forms part of the limbic system, which is responsible for the brain's emotional responses. The term "limbic" comes from the Greek *limbus*, meaning margin, border or edge, which reflects the French surgeon Paul Broca's original observation of the region circling the inner margin of the hemispheres. The amygdala is the centre for basic feelings, particularly fear and sexual responses, and receives inputs from the visual, auditory and sensory parts of the cortex, with numerous further connections to the frontal lobes. Thus it has a crucial and important role in the mediation and control of major emotions like friendship, love, rage and aggression; emotions that are fundamental for self-preservation. The output from the amygdala goes to the *autonomic nervous system*, which, through chemicals such as adrenaline, brings the body to a state of alertness. Destruction of both amygdala essentially tames an animal; it becomes sexually non-discriminative, devoid of affection and indifferent to danger, whereas electrical stimulation of these regions elicits violent aggression. Humans with marked lesions of the amygdala fail to connect emotional to perceived information. For example, on seeing a familiar person, they know who the person is, but are not capable of deciding whether they like or dislike them.

The hippocampus is where memories are stored and this includes the emotional processing that accompanies them. One can see how the hippocampus might be involved in comparing the conditions of a present threat to those experienced in the past, in order to formulate the best response. The hippocampus is therefore an integral part of the limbic system, combining episodic memories with information from the amygdala and other limbic structures.

Another component of the limbic system is the part of the brain responsible for processing smell, the *primary olfactory cortex* in the temporal lobe. This is some distance from the actual nerves concerned with detecting smell – the olfactory bulbs, which are located at the base of the frontal lobes. The path between the two is a very primitive route via the *entorhinal cortex*. This is rather vestigial in humans and is composed of only four layers, at the expense of the more recent evolutionary development of the six-layered complex *neocortex* in the much larger hemispheres. In hedgehogs, which have an exquisite sense of smell, the entorhinal cortex forms the major part of the brain. The speed and strength of the connection between smell and memory can be exploited and is one of the reasons it can be a good trick to try a new perfume or aftershave before going on a special holiday. By doing this, happy memories will be reinforced and triggered by the same smell in the future. Often the memories are far more vivid than those produced by photographs. This also ties in with the fact that some people with temporal lobe epilepsy report smelling unpleasant smells during auras before seizures, confirming that the temporal lobe is intimately involved in the processing of this sense.

thalamus – gathering information

Despite being rather less nose-driven than our prickly insectivore friends, the association of smells with intense memories is something we all recognize. All sensory information on its way to the six-layered neocortex has first to pass through a central "relay" and processing structure, the *thalamus*, from which it is directed to various cortical regions. The thalamus sits centrally in the brain, so much so that it lies next to its partner in the other hemisphere. It acts as a relay station for sensations, receiving virtually all the signals arriving in the brain and sending them on to their correct destinations. Because of its commanding position, it has, in the past, been thought to be the seat of consciousness. Damage to the thalamus, for example by stroke, can result in severe pain syndromes that are very resistant to treatment. However, smells take a far quicker route, bypassing the thalamus to plug directly into the limbic system.

hypothalamus – the hormone autopilot

Another critical part of the limbic system, tucked deep in the centre of the brain, is the *hypothalamus*. This is a very ancient structure in evolutionary terms and has several crucial functions, including the regulation of body temperature, food and water intake, patterns of sexual behaviour, fear and rage, reward and punishment, and sleep–wake cycles. It has connections with all other parts of the limbic system and also with the *brainstem reticular formation*, as well as receiving hormonal signals from all over the body. The hypothalamus gives rise to the *pituitary gland*, which hangs below it at the base of the brain. Through both inhibiting and stimulating signals from the hypothalamus, the pituitary is responsible for the release of several chemical messengers including sex hormones, growth hormone, thyroid stimulating hormone, a steroid releasing hormone, prolactin (involved in breast-feeding) and an anti-diuretic (water-conserving) hormone that acts on the kidneys.

basal ganglia – oiling the wheels of movement

Deep within the brain, towards its base, is a series of grey matter structures, the *basal ganglia*, whose components include the *caudate*, *putamen* and *globus pallidus*. The basal ganglia have a critical role in motor function, acting on a large scale. The major neurotransmitter substance involved in this role is dopamine. It is produced in a part of the brain that appears darker then the surrounding tissue and is therefore termed the *substantia nigra* (black substance). The degeneration of this part of the brain was first described in 1817 by the British physician James Parkinson (1755–1828). Parkinson's disease is characterized by slowness of movements, often with difficulty in starting off. The normally smooth contraction and relaxation of muscles is lost and generalized rigidity occurs. Imbalances in the control of muscle movements can also lead to tremor. It is possible to treat Parkinson's disease by giving the drug L-dopa, which is able to cross into the brain, where it is converted to dopamine, boosting the levels and restoring movement. There are also more radical treatments, involving surgery to destroy or stimulate small parts of the basal ganglia, in an attempt to restore balance to the system.

cerebellum – a fine balancing act

Cerebellum means "little brain". The Greek physician Erasistratus concluded that the cerebellum must be the location of the soul. He understood what we now know to be its fundamental role in the co-ordination of fine movements. As the motor system of mammals became more sophisticated, there was a need to co-ordinate increasingly accurate movements, such as those of the eyes, hands and fingers, often while balancing precariously and moving rapidly. These increasing demands on the cerebellum have resulted in its evolution and enlargement. This is evident in its structure, in which the central part is the oldest and most primitive, and the outer parts of each lobe are concerned with functions unique to humans. There are strong connections (inputs and outputs) with the motor regions of the cortex, as well as with the balance centres (vestibular system) of the inner ears.

The cerebellum has two lobes, but, in contrast to the motor cortex of the brain, each lobe controls movements on the same side of the body as itself: left cerebellum co-ordinates left face and body, and vice versa. When the cerebellum goes wrong, this results in impaired co-ordination, or *ataxia*. The commonest cause of acquired cerebellar ataxia in this country is fortunately usually a temporary one – alcohol. The results of this ataxia are familiar: staggering with a broad-based gait, clumsy gestures, slurring of speech and "rolling" of the eyes (really jerking movements). Ataxia can occur as part of cerebral disease as well, for example in the inflammatory condition of multiple sclerosis, in which neurological examination reveals poor co-ordination and jerky eye movements (nystagmus). In cerebellar disease nystagmus may occur simply on movement of the eyes to the side, or even at rest, and can be very disabling.

The cerebellum is separated from the rest of the brain above by a tough fibrous sheet, the *tentorium*. The space below this sheet is not very large, which means the pressure can rise very quickly if there is swelling or a tumour. Severe swelling can occur after a stroke in the cerebellum and is one of the rare circumstances where it may be necessary for a neurosurgeon to relieve the pressure by temporarily removing part of the skull.

NYSTAGMUS – JERKING EYES

Nystagmus is the rapid flicking of eyes, when trying to fixate on an object. It is caused by an initial successful fix, followed by a gradual drift away, usually because the brain somehow believes the head is moving. The balance mechanism corrects the fixation of the eyes automatically, compensating for the illusionary movement. It eventually becomes obvious that fixation has been lost and a rapid corrective movement is made to fix successfully on the object again. This occurs repeatedly, resulting in rapid jerking movements of the eyes.

Nystagmus also occurs naturally, in certain circumstances. One of the best ways to observe it is to watch someone looking out of the window on a train as it enters or leaves a station. (The London Underground is particularly good for this.) At a critical speed the passenger opposite may become fixated on, for example, a platform poster. Their eyes remain fixed on it until fixation is lost because the train has moved on. They then jerk back to the midline to fix on a new object and the cycle repeats. This is *optokinetic nystagmus* and is a reflex that can be used to test for blindness, even in infants or people otherwise unable to communicate.

brainstem – the bridge to the outside world

The brainstem receives all the sensory inputs from around the body, information such as pain and joint position sense, and the motor output from the cortex. These pathways are grouped in discrete, tightly organized bundles of nerve fibres. It is divided into three parts, which are, from the top down, the *midbrain*, *pons* and *medulla*, below which it passes through the base of the skull to join the spinal cord.

midbrain

This is the point at which the large outbound motor pathways, the *corticospinal tracts*, merge into the brainstem, one column from each side. It is also the location of the substantia nigra (see page 57). There is also a very important group of nerve cells, which are the final stage of the system controlling eye movements. We have already discussed the importance of eye movement co-ordination with other systems but, crucially, the eyes must co-ordinate with each other; otherwise double vision will ensue. Specific nerve pathways link the movements of the two eyeballs together and these can become disrupted in certain diseases of the brainstem such as strokes, multiple sclerosis and brain tumours.

pons

The pons (from the Latin meaning "bridge") spans the gap between the midbrain and the medulla and has a distinctive corrugated surface. As well as the sensory and motor pathways, bundles of nerve fibres from the cerebellum merge with the brainstem in this area. There are also centres here for the cranial nerves supplying the face with sensation and movement.

medulla

The final part of the brainstem is the *medulla oblongata* (from the Latin for "marrow"). It contains the nerve cell bodies of the cranial nerves which control swallowing and tongue movements and is the point where most of the pathways for movement and sensation of the limbs cross over to the opposite side of the body.

A network of nerve fibres, the *reticular formation*, also runs through the back of the brainstem at this point. It samples the information carried by most sensory, motor and autonomic nervous system pathways. The reticular formation uses some of this information in life-maintaining reflexes, such as blood pressure, heart rate, breathing,

SUDDENLY LOCKED INTO A COMPLETELY PARALYSED BODY

Locked-in syndrome refers to the, fortunately rare, neurological disorder whereby there is complete paralysis of all voluntary muscles in all parts of the body, except for those that control the ability to look up. Individuals with locked-in syndrome are conscious and able to understand what is going on around them, but unable to speak or move except to look upwards and sometimes to blink. It can occur suddenly and the trapped individual may not be aware that they are able to look up. If medical staff are unaware of the condition, the locked-in person may be mistaken for someone who is unconscious or brain dead, even though they are in fact completely aware and able to feel everything. Fortunately, formal tests for brain death, which are legally required before it can be declared, will reveal the locked-in syndrome. This extremely distressing state can result from a variety of causes but a stroke in the pons is a common one.

swallowing and coughing. It sends outputs downwards towards the spinal cord to affect some aspects of movement, control the sensitivity of spinal reflexes and regulate the how much sensory information, especially pain, is allowed up into the brain. It also sends outputs upwards from the *reticular activating system*, which affect the level of consciousness and so it is important in sleep–wake cycles (see chapter 10).

the autonomic nervous system – a "thought-less" development

The autonomic nervous system concerns all the functions we take completely for granted, that is, those which require no voluntary input or conscious thought to function; for example, breathing, heart rate, blood pressure and digestion. It is the body's autopilot, which allows the rest of the brain to deal with far more challenging problems, such as which television channel to watch. The autonomic nervous system has its own nerve fibres, which permeate the body, often "hitching" a ride alongside other fibre pathways, and its own system of chemical transmitters, including adrenaline. It receives input from the hypothalamus and limbic system, as well as from receptors tuned to the blood level of carbon dioxide (the waste product of breathing), which can then be used, for example, to increase the breathing rate.

We can split the autonomic nervous system into two sections – the *sympathetic* and the *parasympathetic*. The sympathetic system is all about preparing the body for "flight" or "fight", for which adrenaline is the principal chemical transmitter. Its effects include quickening the heart and breathing rates, narrowing blood vessel diameter to increase blood pressure and diverting blood away from digestion to the muscles. In contrast, the parasympathetic system is all about conservation of energy resources, the promotion of blood flow to the digestive system to absorb energy from food, and the slowing of heart and breathing rates. Thus, the two halves of the autonomic nervous system oppose each other and at any moment are finely balanced.

Having a collection of nerves that are so important means they need to be protected from harm, kept well and happy and given everything they need. In the next chapter we will look at how this happens.

the supporting structures of the brain – what a brain needs to survive

Don't lose you head,
To save a minute,
You need your head,
Your brain is in it.

Burma-Shave 1950s roadside advertisement

the brain's supporting cast

Before we look in more detail at the various highly complex functions of the brain itself, we must consider the supporting systems that make them possible; the skull, the cerebrospinal fluid and ventricular system, the meninges and the blood supply.

the skull – the outer packaging

The human brain is the most public organ on the face of the earth, open to everything, sending out messages to everything. To be sure, it is hidden away in bone and conducts internal affairs in secrecy, but virtually all the business is the direct result of thinking that has already occurred in other minds.

Lewis Thomas ("Notes of a biology watcher: on probability and possibility")

The brain is undoubtedly quite delicate, but has no pain receptors itself, so can be pressed, squashed or cut without pain. The apparent insensitivity of the cerebral cortex to direct mechanical and chemical stimulation was, until the nineteenth century, used as an argument against it having any important functions. There are certainly large reserves (as in other organs, such as the kidney and liver). This is just as well, given the many circumstances in which a significant number of brain cells can be damaged – for example, each time a footballer heads the ball. These reserves are not endless, as we can see clearly in some retired professional boxers who have developed dementia or a Parkinson's-disease-like condition, caused by loss of too many neurons. Apart from the risk of these diffuse brain injuries, there is also the risk of direct damage to a particular region of the brain and subsequent loss of function, be it paralysis, loss of speech or blindness. This means we need a rigid casing for trauma protection, which is what the skull provides.

The skull is made up of several bones which come together along *suture lines*. These zigzagging lines are visible on the surface of the adult skull. During embryonic development the skull bones float next to one another. This is crucial, to allow for the growth of the brain and for the squeezing of the head during the passage through the birth canal, when the skull is subjected to extreme forces. After birth, two gaps remain over the top of the skull. The smaller, *posterior fontanelle* is towards the back of the head; the larger, *anterior fontanelle*, is over the front of the top of the head. (The name "fontanelle" comes from the Latin *fons* meaning "fountain", because the pulsing of the brain with each heartbeat can be felt here.) The fontanelles finally close up at around eighteen months of age, followed quickly by the fusion of the skull bones. Then, the only opening is in the base of the skull – the *foramen magnum*, which allows the passage of the spinal cord carrying the bundles of nerve fibres out to the limbs and body organs. There are also several other, much smaller, holes symmetrically placed further forward in the base of skull, through which the cranial nerves leave the skull. The inside of the skull is divided, for anatomical purposes, into three sections – the *anterior fossa*, *middle fossa* and *posterior fossa* (from the Latin *fossa* for a "ditch").

The bones of the skull are not of uniform thickness, but this has no relation to the underlying abilities of the brain. Evolution has struck the balance between a skull that is strong enough to be protective and one light enough to be useful. This means that the design is optimized to prevent the commonest types of head injury from causing brain damage, but may not be so useful against other types. The back and front of the skull is relatively thick but the sides, overlying the temporal lobes of the brain, are somewhat thinner and vulnerable to trauma. Moreover, bleeding from the blood vessels travelling through the skull at these points can produce rapid pressure damage to the underlying brain, the very structure the skull was designed to protect. Although affording excellent pro-

RAISED INTRACRANIAL PRESSURE

Large strokes can cause swelling of the brain, which raises the pressure inside the skull. Certain infections of the brain can do the same. Expansion of tumours within and around the brain will compress the softer brain in preference to the hard inflexible skull and neurological damage will result. Brain tumours are among the commonest important causes of raised intracranial pressure. The initial manifestation of this raised pressure may be a headache (though it must be stressed that the reverse is not true – the vast majority of headaches are not due to raised pressure inside the skull). As time goes on, the raised pressure is reflected in swelling of the optic nerves. In its later stages this can be seen by opticians or neurologists using an ophthalmoscope.

Here is one of the first descriptions, in 1614, of a tumour affecting the brain lining. It is a type called a *meningioma*. The description is by Felix Platter (1536–1614) from *Observationes in Hominis Affectibus* (cited in *Bulletin of the History of Medicine*, 1956):

> There was discovered on [the corpus callosum of] the brain a remarkable round fleshy tumour like an acorn. It was hard and full of holes and was as large as a medium-sized apple. It was covered with its own membrane and was entwined with veins . . . We perceived that this ball by compressing the brain and its ducts with its mass and by flooding them, had been the occasion of the lethargy and listlessness and finally of death.

If the pressure rises very quickly then individual cranial nerves can become pressed against the internal bones of the skull, producing neurological problems, most commonly affecting the pupils or eye movements. If the pressure continues to rise, the entire brain is squeezed out through the only exit – the foramen magnum. In this case, the first structures to be compressed are the vital respiratory centres of the lower brainstem and death quickly follows. Fortunately, modern medical and surgical techniques exist that can sometimes help to dramatically reduce brain swelling before vital brainstem functions are affected.

tection from external trauma, a sealed, rigid casing means that any increase in the internal volume (as from hæmorrhage), must result in increased pressure and subsequent brain damage.

The average adult human head (skull and brain) weighs 4.5 to 5 kilograms (10 or 11 pounds). To maintain the head in an upright position and allow it to move, large groups of muscles are anchored to the surface of the skull. Some of the largest and most powerful of these come up from the back, neck and shoulders and are attached to the skull bones at the back of the head. These are the muscles that are thought by some to give rise to tension headache. This sort of headache unfortunately does not respond very well to painkillers and, in fact, taking these in large quantities can cause a headache in itself. Simple measures such as massage and improved posture are sometimes enough, but occasionally specialized drugs are needed. In rare cases, and for largely unknown reasons, these muscles can become permanently and abnormally contracted. In severe cases this can cause the head to permanently pull towards one side in a condition known as *torticollis*.

csf system – water on the brain

Just as one wouldn't send a valuable fragile item in a cardboard box without protecting it with bubble wrap, the brain is cushioned in a liquid, *cerebrospinal fluid* (CSF). This provides an effective shock-absorption system for all but the heaviest of blows to the head. Interestingly, it is far less effective at dealing with acceleration injuries, such as those experienced in car crashes, where the majority of the damage to the brain tends to occur on the opposite side to the force – the so-called "contra-coup" injury. Recently, it has been suggested that the relatively reduced density of the brain, compared with the surrounding CSF, is responsible for this phenomenon. Perhaps humans are evolutionarily well adapted to deal with everyday knocks or even aggressive blows to the head from the blunt weapons of our enemies, but not, as yet, prepared for the dangers of high-speed travel.

Cerebrospinal fluid extends down around the brainstem and through the hole in the base of the skull, bathing the spinal cord all the way to its termination in the lower back. Around half a litre of CSF is produced per day, deep within the brain in large cavities, or ventricles, from an area known as the *choroid plexus*. Having largely been produced in the lateral ventricles, it flows forward through the third and fourth ventricles before doubling back and out from the brainstem, down around the spinal cord and back up around the top of the brain, where it is reabsorbed into the large veins running over the surface.

Normal breathing and circulation are believed to help the flow and mixing of the CSF. With age and the natural shrinking of the brain (accelerated in certain types of dementia), the CSF spaces around the brain and the ventricles tend to enlarge. If, for some reason, a blockage forms in one of the narrow passages connecting the ventricles of the brain to one another, or not enough CSF is absorbed into the draining veins, an excess can build up. Depending on the site of the blockage, this can lead to enlargement of the ventricles. This is *hydrocephalus*, literally "water on the brain" and can arise through tumours, after hæmorrhage or as part of an infection.

The main role of CSF is to provide a stable and optimal environment for brain cells, allowing efficient transport of chemicals across their membranes. A highly specialized barrier exists between the blood supply to the brain and the CSF. This is the *blood–brain barrier*, which regulates the flow of natural chemicals. This maintains equilibrium and also prevents the brain being exposed to toxic substances (which includes drugs) that might have managed to enter the bloodstream – which means the treatment of brain diseases using medicines can be challenging. Healthy CSF is clear and colourless, like water, even though it has a role similar to blood. CSF can provide information on the chemical status of the central nervous system, as it can be extracted easily using a needle inserted into the spine below the level of the spinal cord; a procedure known as a *spinal tap* or *lumbar puncture*. Laboratory analysis of the fluid can be an important first clue to infections, certain types of brain hæmorrhage and inflammatory diseases of the central nervous system.

the meninges – the brain's clingfilm

The specialized linings of the brain, the meninges, are closely attached to the inside of the skull as a tough, very tightly applied sheet, the *dura mater* ("hard layer"), and contoured around the brain itself the softer *pia mater*. Between them lies a layer of spidery vessels and tissues, the *arachnoid mater*. These three linings are an additional layer of protection for the brain and also carry blood vessels and pain nerve fibres. The brain itself has no pain fibres; it is only once the meninges become inflamed (meningitis) that pain is felt. The result is headache and neck stiffness, worsened by stretching these layers – for example by trying to place the chin on the chest or raising the legs while lying down (a common test by doctors). Bacterial infections of the meninges, such as those caused by the *Meningococcus* bacterium, can be life-threatening and require urgent antibiotic therapy. Viral infections, for example flu, are much more common causes of mild meningitis and usually clear without treatment.

the blood vessels – the fuel supply system

Although the brain is master of the organs of the body . . . it is not so placed that it can survive or have any power in the absence of their help. . . . On the contrary, the animal spirits, and life itself, are so dependent on the continuous supply of blood to the brain, that every . . . suppression . . . soon leads to syncope and unconsciousness, and, further, if such processes persist unduly long, the life ceases completely.

Richard Lower (physician, 1631–1691)

In proportion to our body mass, our brain is three times as large as that of our nearest relatives. This huge organ is dangerous and painful to give birth to, expensive to build and, in a resting human, uses about twenty per cent of the body's energy even though it is just two per cent of the body's weight. There must be some reason for all this evolutionary expense.

Susan Blackmore (psychologist)

Estimated amount of glucose used by an adult human brain each day, expressed in M&M's®: 250

Harper's Index

Every tissue in the body needs blood, to supply oxygen and glucose and to remove waste products. The brain is no exception: indeed it requires around fifteen per cent of the output of oxygenated blood from the heart. If the flow drops suddenly, for example because the blood pressure suddenly drops, the brain is forced to shut down, leading to loss of consciousness. The brain is normally very good at maintaining its blood supply over a wide range of blood pressures. Unfortunately, there are times when the blood supply to a part of the brain is interrupted catastrophically, leading to damage. This is a *stroke*, a term derived from the idea of a "Stroke of God". The interruption in the blood supply results in oxygen starvation (called *ischaemia*) and neurological damage. In the majority of cases, this blockage in the blood supply arises as a result of a blood clot that has either arisen in the area because of "hardening of the arteries" or been swept up from a diseased vessel further away. Such clots eventually disperse themselves, but the time taken determines the extent of the damage to the brain area supplied by that vessel. This is why people susceptible to these sorts of strokes are given blood-thinning drugs such as aspirin or warfarin to prevent future clots.

More recently, attention has been turning to the possibility of using powerful clot-busting drugs similar to those used in heart attacks, and use of the term "brain attack" is being encouraged to give the condition the same urgency in people's minds.

Stroke can, more rarely, be the result of a hæmorrhage deep within the brain tissue itself. More commonly though, a hæmorrhage within the brain arises from an abnormal expansion of an artery, usually at the point where it divides into smaller vessels. These expansions, or *aneurysms*, are a ballooning out of the vessel walls. The artery wall then becomes thin and fragile, bursting relatively easily. Aneurysms can be present from birth without causing any problems; they can also occasionally give warning symptoms before bursting. The result of an aneurysmal hæmorrhage can be severe; sufferers usually have a very sudden, severe, instantaneous headache – some people describe it as like being hit on the head by a concrete block. If the sufferer survives, they may have an operation to clip or seal any remaining aneurysms.

The results of a stroke are very variable and depend on which part of the brain is affected. The pattern of functions affected after a stroke enables neurologists accurately to localize the area involved, so that modern imaging techniques can then home in on this region to confirm the diagnosis.

We have seen how the brain is made, arranged and kept functioning. It takes years for it to mature, a process not complete until we are in our early twenties, but what about what the brain does? Do we learn how to behave or think or are we born with the ability?

the development of behaviour and reasoning – learning to be human

From the moment of birth, when the stone-age baby confronts the twentieth-century mother, the baby is subjected to these forces of violence called love, as its mother and father have been, and their parents and their parents before them. These forces are mainly concerned with destroying most of its potentialities. This enterprise is on the whole successful.

R.D. Laing (psychiatrist and author, 1927–1989)

programmed from birth

Are we born with the knowledge of how to behave or react to our surroundings? It might seem ridiculous that we are programmed to react a particular way at birth, but we know that for some behaviours this is true. We will see later that it is possible that it is true for most behaviours, even though we may not realize it. For new-borns, reflexes lead the way for their first reactions to the environment. Newly hatched birds open their mouths to feed, new-born foals stand and run and new-born human babies turn their heads to suckle; but what about more complex behaviour? Can this also be programmed? Some years ago, the biologist Konrad Lorenz showed that goslings will follow the first large, moving object they see after hatching. This *imprinting* has a restricted window of time in which it can occur. For goslings, this is the first twenty-four hours, after which the imprinting circuits are permanently fixed. Goslings usually imprint on their mother because she is the first large, moving object they come across, but will imprint on any large object, so long as it moves. Lorenz showed that they imprinted on cars, on same sex geese, on himself and even on light aircraft. The imprinted object forms the basis for choice of mate in adult life, so this is an example of a programmed response leading to a specific behaviour years later. Imprinting can work both ways, so that parents also imprint on their offspring. For example, mother sheep imprint on the smell of their offspring in the first few hours after birth.

One set of experiments performed in the 1950s by Hartow, which would be unlikely to be approved today, on ethical grounds, showed

PRIMITIVE REFLEXES

A baby is born with ready-to-go, wired-in, reflexes. Touch a baby's cheek and it will turn its head and move its mouth, ready to suckle – the so-called *rooting reflex*. Stand a baby up and its feet will push against the ground with an antigravity reflex. Push its feet gently against a step and it will step up. Put your fingers in its hands and it will grip tightly on. Even in water, a baby has a diving reflex that slows the heart and prevents it breathing until it is safe to do so. What happens to these primitive reflexes? As we develop, they are inhibited by the frontal lobes so that by the time we are adults they have largely disappeared (although in various studies, up to twenty per cent of adults still have primitive reflexes of some kind).

People who sustain damage to the frontal lobes, for example through a stroke, unmask the primitive reflexes, which can be revealed by neurological examination. Scratching the outer edge of the sole of the foot in an adult with a normal nervous system results in the big toe pointing down. In someone with damage to the frontal lobes it instead produces the first part of the antigravity withdrawal reflex (the Babinski sign) and the big toe points up. Other primitive reflexes revealed are to scratch the palm of the hand from base to below the index finger and the chin on the same side will wrinkle (part of the hand-to-mouth feeding reflex) Stroke the cheek and a rooting reflex will follow. Tap the lips and the lips will pout, ready to suckle. Stroke the palm vigorously with your fingers and the grasp reflex will prevent release of your fingers.

The diving reflex is not inhibited by the frontal lobes and there-fore persists in normal adults. This makes it useful medically. In some people with an abnormally fast heart rate, placing their face in a bowl of cold water switches on the diving reflex and slows the heart down, correcting the abnormal rhythm, which remains reset even after they resurface.

that infant monkeys are born with an expectation of what a mother should feel like. Infant monkeys given a choice between cuddling a "mother" consisting of a bare wire frame or a wire frame covered with cloth chose the cloth-covered frame, even when the wire frame had milk and the cloth-covered frame did not. The comfort from the cloth outweighed any comfort from feeding. Some reactions in humans are hardwired too. Experiments in humans and monkeys have shown that there is an automatic fear reaction to snakes or to the sudden approach of a large object (looming), even without previously being exposed to such things.

It seems that we are born with at least some built-in behaviours and responses. Some people have called this *phylogenetic memory*, which means a memory that is programmed into the nervous system of a species. It is not a memory that results from learning, but one which has been collected through natural selection and evolution, so that we are born with the knowledge, just as if we had learnt it. Presumably, this means that the neural circuits for this behaviour are, from the start, wired up as they would be if the behaviour had been learnt. A less extreme version of this is what gives us a brain designed the way it is. Each part of the brain has a tendency to respond to certain things in certain ways but our experiences determine the detail. Most of our reactions are learnt step-by-step in a very long process, as we shall see.

learning to be human

What counts as normal behaviour for any given situation depends on our society and culture. It also depends on our age, sometimes on our sex and on our perceived status within society. How, despite a seemingly endless set of social rules, do most of us say and do the right thing most of the time?

The frontal lobes play the greatest part in controlling our behaviour. This is not to say that other parts of the brain do nothing. Obviously we need to perceive the world appropriately in order to respond appropriately. If, for example, we see threats where there are none, or do not recognize real threats, then regardless of how good the frontal lobes are, our behaviour will be inappropriate. The frontal lobes are the largest part of the brain but are relatively neglected by researchers. This is because damage to the frontal lobes may lead to problems that can be quite subtle and hard to measure, such as changes in personality.

The frontal lobes are also involved in speech output, control of eye movements and making and planning movements. These are all covered elsewhere. Here, we will be concentrating on the part of the frontal lobes which control behaviour and social skills, which is at the very front, and the part controlling thinking, which lies more at the sides.

theory of mind – knowing others

> Any fool can tell the truth, but it requires a man of some sense to know how to lie well.
>
> Samuel Butler (composer and satirical author, 1835–1902)

Social skills require an ability to understand others. Animals signal to each other using body postures. When one animal displays, another animal will respond to the behaviour. For many animals, this is an automatic response, without any learning involved, equivalent to humans' ability to smile in response to another smile. Human beings do more than just respond automatically to another person's behaviour – we respond to each other's mental states. We build a model of the other person's beliefs, desires, knowledge and intentions and respond to this, rather than to their behaviour. We interpret their behaviour in the light of our internal mental model, so that the same behaviour from different people in different situations leads to different interpretations and responses. This process is slow and painstaking and reaches maturity only after our teenage years. Our ability to build a model of another person's mind means we must each be able to understand the concept of a mind. This is the *theory of mind*, a corner-stone of psychological research into social skills. Without it, we could not empathize with another human being. It is also pivotal in any discussion of self-awareness, since the concept of a mind applies to our own minds too.

understanding the world

The stages passed through as we develop a model of the world and others around us were first researched and described by the Swiss scientist, Jean Piaget (1896–1980). He believed that the development of a child's understanding advances in sudden leaps, followed by more gradual change. Each leap happens around a particular age and, until that age, even bright children will be unable to grasp the concepts of the next stage. During each stage, the child uses the mental concepts they currently have to try to understand the world. A new experience that fits with their present model is assimilated and maintains the child's mental equilibrium. A new experience that does not fit requires an adjustment until the model breaks down because the child's level of understanding allows it to appreciate the flaws in its current model.

Gradually, the mental model of the world becomes better and better. Piaget's stages of development are still used today, although there have been some modifications. In particular, it is now clear that some children reach the third stage sooner than previously thought, and that some (perhaps a majority of) normal adults never reach the final stage. A key feature of Piaget's ideas is that no stage can be skipped. They invariably occur in the order described below and are not specific to any culture. This is likely to be because we all develop similar models of the world, based on our pre-programmed learning instincts and so we come to similar crises at similar times, with similar inevitable solutions.

Human beings are highly social: these skills are very complex and require a great deal of brain power. The brain needs to grow into the skills, changing physically as it grows. The growth, changing and maturing of the frontal lobes exactly parallels the development of our view of the world and of our social abilities.

the stages of development

Sensorimotor (0–2 years)

In infants, the frontal lobes are small, reflecting the fact that we need very little in the way of social skills at this stage of life. This is the sensorimotor stage and is how we understand the world from birth until about two years of age. The first major concept is that the world is separate from ourselves. The key feature of this stage is that seeing is believing. Children in this stage show little understanding that an object they cannot see might still exist – easily demonstrated by hiding something the child wants and looking for signs that the child is searching for it. Learning that hidden objects still exist is an intellectual leap made around the age of nine months. We also develop a sense of identity and begin to act with intention, for example deliberately shaking a rattle to make a noise.

Pre-operational (2–7 years)

Children in the pre-operational stage cannot easily understand abstract ideas, but are beginning to understand concrete physical concepts, such as shape or colour and to represent the world with images and language. They cannot easily take another person's point of view, because they have not yet developed a theory of mind. Their view of the world is egocentric – they have a belief that the world revolves around them.

When asked to classify objects, children in this stage will group them under a single feature. For example, given some blocks of various shapes and colours, they will group them into blocks of the same colour (regardless of shape) or blocks of the same shape (regardless of colour). They will not collect groups of blocks that share colour and shape, for example, all red, square blocks. This is because they cannot yet tackle the ideas of multiple overlapping properties of an object and that an object might belong to a sub-set of a larger set.

Concrete operations (7–11 years)
In this stage of development, which continues up to about the age of eleven, children learn to think in a logical way, mainly about concrete objects. They also begin to understand that other people may have a different view and what that view is. This is the first stage in which they develop theory of mind and can therefore become less egocentric. This can be demonstrated by showing the child a scene and asking them to report what the scene looks like for another person. Until they are able to take another person's point of view, children will always describe the scene as they see it, whereas children with theory of mind will correctly describe what the other person sees. It may seem strange to think that it is possible to have a concept of self without a concept of others, but we are talking about a concept of our own mind compared with the mind of others, rather than the concept of ourselves and others as objects.

During this stage, children also make another major intellectual leap. They discover that the properties of objects around them remain the same even when the objects are manipulated in some way. This is known as *conservation of property*. A classic example is the test of conservation of volume. A child is shown a short fat glass containing orange juice which is then poured into a tall thin glass. When asked which glass holds more orange juice, children in this stage realize that the amount of juice is unchanged, whereas children in earlier stages think that the taller glass has more juice. They believe that the properties of the juice can change and are determined by the surroundings of the object. This may seem trivial to adults, but it requires a significant leap of understanding which we need to make for seven different properties: number, length, liquid, mass, weight, area and volume.

To grasp that the properties of an object might remain unchanged we need to understand three concepts. First, we must understand that a material will stay the same if we do not add or take anything away from the material (the concept of "identity"). Second, we have to understand that a change in one dimension can be compensated for by a change in another. Third, we must understand that a change can be undone if we perform the steps backwards.

Formal operations (11 years and older)

This is the final stage of development, if reached, and may be achieved around the age of eleven. Early in this stage there is a return to ego-centricity, possibly because of changes in brain wiring in response to puberty, but this eventually disappears. By this stage, we are able to think about abstract things. We also demonstrate a more scientific approach to the world, generating and testing ideas systematically, and can think about the future. This is the stage when people can become interested in ideological problems. We can understand mathematical problems more easily and can reason logically. We can build an abstract model of a world with properties different from reality and test the model for answers to questions. For example, we could answer the questions, "what if rain fell upwards?", "what would happen if we had gills?" or "what if grass were pink?" In industrialized countries, only about a third of high school graduates can think formally, and many never think formally throughout adulthood. Even those who can do not need to for most of their waking lives, because so much of what we do does not require it.

Piaget believed that biological development automatically drives the change from one stage of development to the next. Current evidence suggests that biological maturity is necessary for progression, but also that the child must be in the right environment. This is particularly true for the formal operational stage, which probably needs quite a high level of challenging, formal education. Other factors also propel us through the stages. Our previous experiences mean our models have been challenged sufficiently for us to need the next set of concepts. Social interaction means that developing a theory of mind is useful and possible. These factors lead to behavioural and intellectual development because of an innate, instinctive need to make sense of the world, which Piaget called "equilibrium". Once experience or social interaction disturbs our equilibrium we must accommodate it by changing our model and establishing a new equilibrium. Clearly, our model must be sophisticated enough to recognize when it is no longer adequate, or our equilibrium would never be disturbed.

evolutionary psychology – designed to think and behave as we do

We started this chapter by asking if we have innate behaviours. We will now try to understand if such programming could explain why we think and behave as we do.

Human expressions are universally understood; even people blind from birth make appropriate expressions, despite never having seen one. Everyone smiles with pleasure and cries with pain. Why are so many of us scared of spiders or heights or public speaking? Why do we greet each other before interacting? Why do people risk their lives for strangers? Evolutionary psychology is a body of theory that attempts to answer these questions by regarding our thoughts and behaviour as a set of responses programmed into us by evolution because they were useful during our prehistoric past.

A fundamental concept of evolutionary psychology is that the brain organizes behaviour, so any questions about why we behave as we do will have an answer, at some level, in the circuits of the brain. Only organisms that move have a brain. Another way of putting this is to say that only organisms that show behaviour have a brain, so a basic function of brains is to generate an appropriate behaviour in response to the environment. The circuits in our brains were not designed to solve any old problem, but to solve specific, life-altering, problems that come up repeatedly and affect our reproductive fitness. For example, knowing what is safe to eat, how to recognize danger, how to respond to poisonous animals and how to respond to great heights. Only these types of problems can be acted on by natural selection, so they are the ones we would expect to be programmed in.

instincts – the programmes we were born with

Most people think that animals are ruled by instincts but that humans can rise above this: that, apart from a few basic instincts like hunger or thirst, we choose what to do and how to do it. This is probably not true – it is likely that we have far more instincts than any other animal. This

IF YOU DON'T NEED YOUR BRAIN, EAT IT . . .

The sea squirt starts out life as a larva with a cerebral ganglion (equivalent to a very primitive brain) that controls its movement and a visceral ganglion (equivalent to a very primitive autonomic nervous system) to control digestion. Once it finds a good rock settle on, it attaches itself permanently and then digests its cerebral ganglion. The main purpose of a brain is to allow an animal to move purposefully, so, to a sea squirt, once it has settled, a brain is just so much baggage.

is because most of our day to day decision-making goes on subconsciously; most of it probably far more complicated than other animals experience. Only the most processed information makes it to our attention. We balance, we make facial expressions, we breathe without thinking about it. We can alter these behaviours if we choose, but they go on automatically if we leave them to it. Every sense is highly processed by our brain circuits, as we will discover later in this book. What seems like a single sense is really a synthesis, presented to our conscious mind as an entity ready to be understood and experienced. For example, the sense of touch is really a compound of light touch, joint position, pressure, pain, temperature, stretch and many other senses. Even these initial senses are used to generate higher senses (such as the two-point discrimination sense that tells us if two pieces of sense information come from a single source or not) and combined to give a seamless sense of feeling. The sense of vision is similar. We can detect motion, edges, colour, size and orientation within the retinal circuits. These are synthesized to generate three-dimensional vision. Even this is processed to identify faces, animals, family, etc. specifically. The central vision is sharp but we have a blind spot in each eye and low acuity vision around the periphery, neither of which bothers us. All of this is integrated into a single visual experience, presented to our conscious mind. The various senses are then seamlessly fused into a unified experience of being, so that we are aware of a particular sensation or experience only if it warrants conscious attention.

These neural circuits do more than simple processing of sensory signals. They can recognize complex circumstances, such as an embarrassing social situation, a potential accident or a relaxing occasion, and activate appropriate responses. This happens subconsciously; we can consciously respond if we wish, but even then it can be difficult to override our pre-programmed responses. Standing up to give a speech or lecture, it is difficult not to feel nervous unless we are practised; picking up a harmless spider can be impossible for someone who has developed a phobia. We have an in-built tendency to develop certain responses, such as phobias of heights and snakes, and we are pre-programmed with others, such as the fight or flight response, because they had a survival advantage for our ancestors.

specialized circuits for behaviour – each jack of one trade, together master of all

This means that what seems like a relatively simple task, such as recognizing a loved one, is the result of a series of highly effective problem-solving circuits that have extracted the relevant information from our environment and resolved it into a single experience, presented to the conscious mind. What is more, our circuit responses are not randomly chosen, but strongly weighted towards being picked from a set of likely solutions. We are wired by evolution to favour particular behaviours and programmed by experience to select them.

Just as a hammer bangs in a nail more efficiently than a saw, we need brain circuits that are specialized to solve some specific problems better than others. Evolutionary psychology regards the brain as a set of mini computers, each designed to solve its own problems, such as extracting information from the eyes, processing information from air vibrations arriving at the ear, choosing a sexual partner, selecting what to eat, etc. Having such specialized circuits is an excellent way to deal with the world quickly and well.

here's the answer, now what's the question?

Computer science recognizes two ways to solve a problem. The first is to design a set of general rules, for example probability theory, which applies equally well to packs of cards, dice, quantum physics and genetics, or calculus, which applies to sailing on rivers, gravity, population growth and radioactive decay. The content of the problem does not matter, just the properties. The second approach is to use a large set of specific assumptions and presumed knowledge. Because the content of the problem is central to the designed solution, the method cannot easily be generalized. We would expect that any problem-solving circuits designed through evolution would be of this content-dependent type. There is plenty of evidence for this in human behaviour. For example, a new-born baby has an in-built tendency to recognize faces. At only a few minutes old, a new-born will preferentially turn to look at a pattern arranged like a face rather than a random pattern. At a few months old, infants expect objects to be solid and think of any whole object that moves as a single entity, regardless of its size, shape or colour (contrast this with a cat, which is programmed to detect and respond to movement, regardless of what the whole object is, and might well attack its own tail or the owner's hand), and are surprised by apparently solid objects passing through each other. Babies less than a year old can distinguish animate from inanimate objects, even when the inanimate objects are moved, and attribute the movement of animate objects to that object's unknown goals and desires. Toddlers assume that an adult making a word-like sound and pointing to an object is referring to the whole object rather than a part of it. These observations suggest that, even early in development, we have certain assumptions. Although it could be argued that reactions occurring months after birth have been learnt, experiments can show that they are not. Just because some of these abilities are not present at birth does not mean they are learnt. A baby does not have teeth at birth but this does not mean we learn to have teeth. These in-built assumptions are what make a young child intelligent and able to learn about its environment quickly. The ability to guess what others are thinking (theory of mind) is particularly important in human development; even though we know that they do not yet have the ability to conceive of another mind, experiments show that toddlers can understand, from eye movements and gaze direction, what other people believe or want.

a multi-faceted mind – our reasoning instincts

The solutions from each specialized circuit are specific to the kind of problem it is designed to deal with. A circuit specialized for vision will not process sound information without producing potentially strange results. A circuit specialized to understand what others are thinking will not necessarily give appropriate answers about inanimate objects. On the other hand, such results may provide an insight into the problem that is not available from the specialized circuit. Also, any problem is likely to be composed of many smaller problems, which means that it can be useful to try to solve problems with several different circuits at the same time and compare results. In other words, different parts of the brain will be activated simultaneously when processing any one piece of information – something we see happening in brain scans.

We often deride instincts, as if they were the opposite of logic. In one sense they are, because logical systems can be generalized to apply to any problem, whereas instincts are specific. This works to our advantage, because we have a huge number of specific problem-solving circuits, each designed to extract information and find a solution to a problem we will definitely face – in other words we have instinctive learning and reasoning circuits. Humans are regarded as "higher" animals, with instincts erased by evolution, dominated by rational thought, but learning and reasoning *are* our instincts. They are as easy for us to use as it is for a bird to fly, or a cat to hunt. The learning and reasoning circuits we develop as children possess all the properties of instincts: completely specialized to deal with particular problems, developing reliably, predictably and automatically in all humans, with no conscious effort and no instructions, applied without awareness of their underlying logic and distinct from our more general abilities to think or behave intelligently.

Evolutionary psychology is an interesting way to think about behaviour and why we are as we are. On a philosophical level it must also make us question how much of what we think we know or do is actually hard-wired rather than free will.

PROGRAMMED TO DETECT CHEATING

Research from leading evolutionary psychologists Leda Cosmides and John Tooby has shown that we have specialized reasoning instincts designed to detect cheating in a social situation. To demonstrate these they used the Wason selection task, a logical test of the "if A then B" variety. For example, we are told that "if someone goes into Boston, they take the subway". We are then presented with four cards representing the behaviour of four people (one person on each card). On one side of each card is their destination and on the other what transport they used. We can only see one side. We are asked to test the statement "if someone goes into Boston they take the subway" by turning over the minimum number of cards. Here is an example: the four cards say "Boston", "Arlington", "subway", "cab". Which cards should be turned over to prove the statement true? These kinds of tests are not intuitive and only about twenty-five per cent of people choose correctly first time. Here, the answer is two cards – "Boston" (because if it does not say "subway" the statement is false) and "cab" (because if it does say "Boston" the statement is false). Cosmides and Tooby then changed the test statement to a type describing a social situation where it might be possible to cheat: "if you are to eat those cookies, you have to make your bed first". The four cards then had statements about people's behaviour with regard to what they ate and what they did first. The test was essentially identical, but the content was different. In this situation, more than eighty per cent of people could answer correctly first time. This was true regardless of their culture, spontaneous and occurred without much obvious prior thinking – the answer seeming to "pop out" at them. This strongly argues for a circuit specialized to detect cheating in a social situation, rather than a general logic circuit for solving "if A then B" type statements that in themselves do not have any survival advantage.

STANLEY MILGRAM AND THE VOICE OF AUTHORITY

A substantial proportion of people do what they are told to do, irrespective of the content of the act, and without limitations of conscience, so long as they perceive that the command comes from a legitimate authority . . . This is perhaps the most fundamental lesson of our study: ordinary people, simply doing their jobs, and without any particular hostility on their part, can become agents in a terrible destructive process.

Stanley Milgram (*The Man Who Shocked the World*).

The psychologist Stanley Milgram performed a series of brilliant and controversial experiments, which showed that we are all programmed to obey authority blindly and only a few of us can say no.

He set up an experiment in which ordinary people were invited by newspaper advertisement to take part in a study on memory and learning for $4.50 per hour. They were met by a stern man in a white coat (the experimenter) and a friendly co-subject. The experimenter explained that they would be studying the effects of punishment on learning. One of them would be the teacher and one the learner. They drew rigged lots, so that the person answering the advertisement always became the teacher. The learner was taken to a room behind a silvered glass panel, strapped into a chair and wired up to an electrode. The teacher was taken to the room on the other side of the glass panel so that they could see the learner. This room contained a generator with thirty switches in 15 V increments, with the voltage ranging from 15 to 450 V. These were clearly labelled from "slight shock" to "danger: severe shock", with the highest two just marked "XXX". Depressing switches activated lights, buzzing sounds and meters on the machine. The experiment involved the teacher reading a list of word pairs to the learner, who had to read them back. If the learner made a mistake, the teacher had to deliver an electric shock, increasing the voltage with each mistake.

What the teacher did not know was that the learner on the other side of the screen was an actor, wired up to nothing. The study was designed so that the learner would make mistakes in a planned way, receiving increasingly severe shocks. With each shock, the learner would scream and writhe in agony. If the teacher questioned the experiment, the experimenter would say, "please continue". The teacher was given a real 45 V shock to know how painful it was.

Continued

Video footage of the experiment shows that although those delivering the shocks were often highly distressed, tearful and reluctant to proceed because of the pain they were causing to the learner, they still continued. The bland statement "please continue" was usually sufficient, with teachers delivering apparently lethal shocks despite being very distressed. This study showed that ordinary people could commit acts of cruelty previously thought only possible for sadistic monsters on the fringes of society. Two-thirds of the subjects were categorized as "obedient" and came from all walks of life: working, managerial and professional classes, male and female. No subject stopped before delivering 300 V and sixty-five per cent delivered an apparently lethal 450 V. Women and men were equally likely to be obedient, although women tended to be more nervous. These experiments have been repeated in many forms and in many societies, always with similar results.

Stanley Milgram reported, "I observed a mature and initially poised businessman enter the laboratory smiling and confident. Within twenty minutes he was reduced to a twittering, stuttering wreck who was rapidly approaching nervous collapse. He constantly pulled on his earlobe and twisted his hands. At one point he pushed his fist into his forehead and muttered, 'Oh God, let's stop it'. And yet he continued to respond to every word of the experimenter, and obeyed to the end."

having and using a brain as an adult

consciousness – the ultimate mystery

Brain: an apparatus with which we think we think.

Ambrose Bierce (author and satirist, 1842–1914)

Five senses; an incurably abstract intellect; a haphazardly selective memory; a set of preconceptions and assumptions so numerous that I can never examine more than a minority of them – never become conscious of them all. How much of total reality can such an apparatus let through?

C.S. Lewis (author, 1898–1963)

the ultimate mystery

In talking about consciousness we inevitably enter a minefield of confused definitions and philosophy. Most of this is because everyone thinks they know what is meant by "consciousness" until they are pressed to explain.

Consciousness can be used to mean wakefulness or a state of arousal. It is clear that when we are asleep we experience dreams and when we are awake many things are going on that we do not experience. For our purposes therefore, the state of arousal is different from consciousness. Some people use "consciousness" to mean self-awareness, but this is too narrow a definition. "Consciousness" is also used to mean "mind" but there are many aspects of the mind that are unconscious. "Consciousness" is therefore many things to many people, so we need to be clear about what we will discuss. In this chapter, we will use the most common definition for consciousness: "that which we experience". This includes both our inner states and our experiences of the world.

where does consciousness happen?

Can we point to the place where consciousness happens? Is there a structure for consciousness? The quest to find an answer to this problem has led to the development of several schools of thought. Most people would instinctively say that the brain is the structure which generates consciousness, as would the philosophers of the first schools of thought, the reductionists and the materialists. These philosophies say that consciousness happens where the processing of nerve impulses takes place. We know consciousness is a physical function of the brain, because things that affect the brain affect consciousness, so the brain, which is the place where nerve signals are processed, is the place where consciousness is generated. Therefore, to understand the mind, we need to understand the brain.

The second school of thought is the dualist. In the seventeenth century the French philosopher Rene Descartes (1596–1650) suggested that the world consisted of two components: material components that had existence in space and a thinking component that was conscious but had no existence or location. In this theory, we cannot say where consciousness takes place, because it has no location in the material world. We could, if pushed, say consciousness happens at the interface between the external material world and the thinking world of consciousness. The most likely place for this is the brain, but we cannot know this for certain.

DESCARTES

René Descartes was a philosophical giant whose methods included the method of hyperbolic doubt. In trying to understand the truth about the world, his method consisted of rejecting anything which could be doubted at all. This left just two things that could certainly be said to exist: the doubting and the existence of the doubter (Descartes himself). From this point he then attempted to reconstruct knowledge, never allowing doubt to creep back in. He reasoned that he himself must have the characteristic of thinking and that this thinking thing (mind) must be quite distinct from his body. He came to this conclusion because his mind had a property that his body did not. His mind could not be doubted to exist because it had to exist to do the doubting, whereas his body could be doubted to exist.

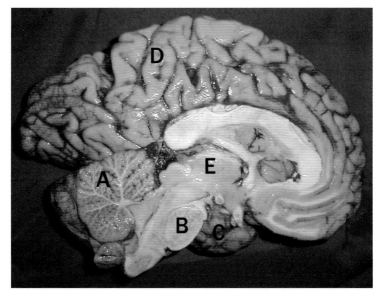

Plate 1 Sagittal section (in the vertical plane straight through the nose and out the back of the head) through the brain showing A: cerebellum, B: pons, C: temporal lobes, D: cerebral hemisphere, E: thalamus.

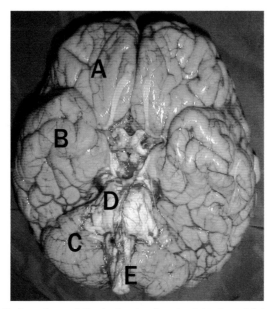

Plate 2 Undersurface of the brain showing A: right frontal lobe, B: right temporal lobe, C: cerebellum, D: pons, E: medulla and spinal cord.

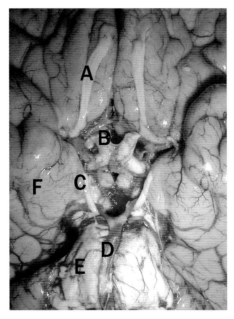

Plate 3 Close up undersurface of brain showing A: right olfactory nerve, B: optic nerves, C: 3rd cranial nerve (moves eyes), D: basilar artery, E: pons, F: right temporal lobe.

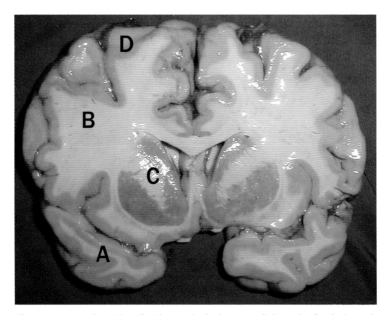

Plate 4 Coronal section (in the vertical plane parallel to the face) through the brain to show A: right temporal lobe, B: white matter nerve fibres, C: the basal ganglia, D: cortex of the right cerebral hemisphere.

The following images are all fMRI scans and consist of a functional component (the coloured regions) and an anatomical component (the underlying brain scan image). By convention, fMRI images are shown so that the right side of the subject is on the viewer's left. This is best imagined for most images by assuming the subject is on their back and the viewer is looking up their nose.

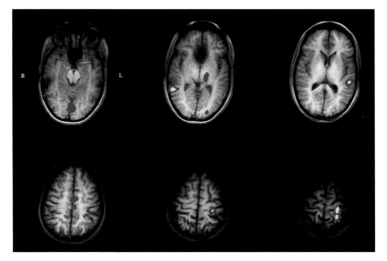

Plate 5 Brain regions activated by 25 single button presses made only in response to specific cues.

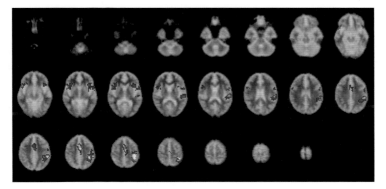

Plate 6 Brain regions activated by moving the right arm.

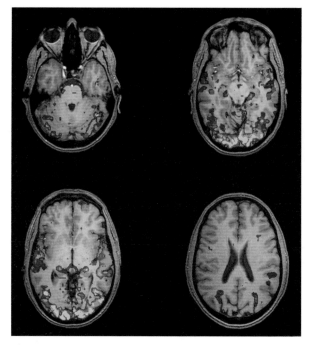

Plate 7

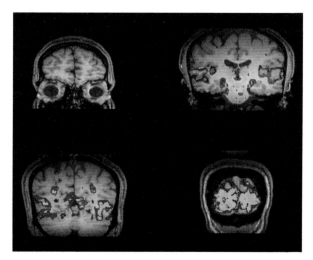

Plate 8

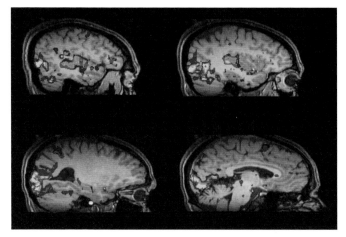

Plate 9

Plates 7, 8 and 9 Functional MRI scans showing brain areas activated in someone experiencing simultaneous audio (blue) and visual (red/yellow) stimulation. Audio stimulation was 45 second bursts of speech, visual was an 8 Hz reversing checkboard pattern in 30 second bursts. Plate 7 shows axial sections (in the horizontal plane for someone standing), Plate 8 coronal sections, and Plate 9 sagittal sections.

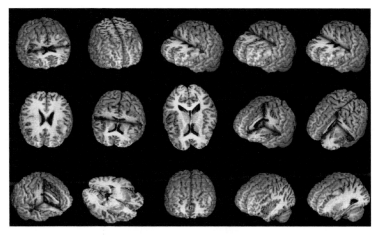

Plate 10 Frontal activation generated by a task to demonstrate the neural activity of creative intelligence.

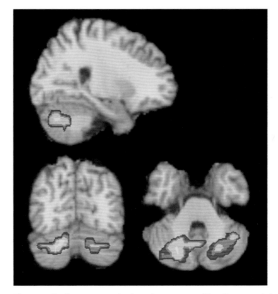

Plate 11 The cerebellum controls hand–eye coordination, and is lit up in this task requiring ocular pursuit with simultaneous control of a cursor using a joystick.

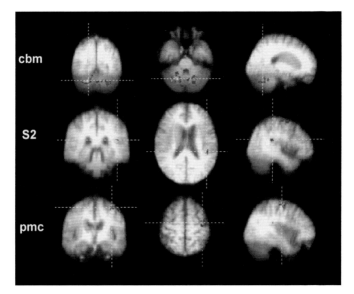

Plate 12 Following a stroke, successful rehabilitation leading to improved hand function correlates with increased activity in the coloured brain regions.

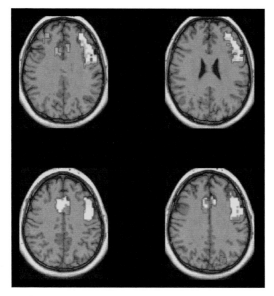

Plate 13 For this scan, subjects were asked to generate a list of words beginning with a particular letter. Several brain regions, including a large region of the dominant hemisphere (left in a right-hander) is lit up. This speech region is Broca's area and is close to the region used for moving the right arm or hand.

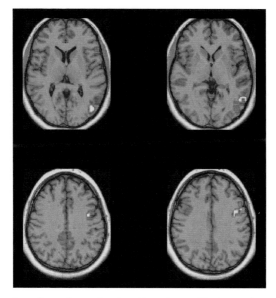

Plate 14 A much smaller region is active when subjects are shown a picture of an animal and asked to name it.

Plate 15 An image reconstructed from MRI scans showing the complete corticospinal tract on one side (red, green, yellow) and part of the tract on the other side (all red).

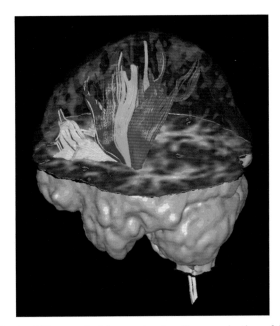

Plate 16 An MRI-generated image showing the organization of individual nerve tracts in the internal capsule, connecting the cortex with the spinal cord.

A third view, the reflexive view, has been suggested by the contemporary philosopher Max Velmans. His idea is that, for example, pain in a finger cannot be said to take place in the brain. Although nerves firing in a particular part of the brain may be part of the process of feeling pain, the pain is actually felt in the body part that is hurt, rather than in the brain itself. Similarly, when we see or hear, the experience is in the external world, not the brain. The experience we have is of events outside our bodies, so consciousness lies somewhere vague and not well-defined, although, unlike in the dualist school of thought, it does have a location.

These three views about consciousness are important from a philosophical point of view, but, from the practical view of a neurologist, what matters is if there is a structure that has specific and predictable effects on consciousness. The brain clearly has these properties and so, for a neurologist, consciousness is generated within the substrate of the brain; therefore, most neurologists belong to the reductionist or materialist schools.

experience – the fundamental particle of consciousness

You can know the name of a bird in all the languages of the world, but when you're finished, you'll know absolutely nothing whatever about the bird . . . So let's look at the bird and see what it's doing – that's what counts. I learned very early the difference between knowing the name of something and knowing something.

Richard Feynman (physicist, 1918–1988)

We have so far discussed one facet of consciousness – which structure is responsible for it – but there are two other important aspects. The first is the objective functioning of the brain: the physical processes that allow us to react to the outside and inside world. The second is perhaps the real mystery: how do the physical processes lead to a subjective experience? How do we know what the colour blue should look like or how softness should feel? The objective functioning of the brain is like a neural network correctly identifying colours, but this is quite different from the subjective experience of experiencing colour. We can understand the electronic circuits that lead a computer to identify colours correctly, but most people do not believe that the computer experiences colour. Even if we understood the brain circuits that lead us to identify colours, we would not necessarily be any closer to knowing why we experience colours in a particular way.

Frank Jackson's thought experiment

To illustrate the difference between identifying and experiencing colours, the philosopher Frank Jackson proposed a thought experiment: imagine that in the twenty-third century, a neuroscientist, Mary, is the world leader in understanding colour vision. She knows how different wavelengths of light correspond to the names of different colours. She understands the functioning of the eye, the nerves that connect it to the brain and all the brain circuits that react to colour. However, Mary lives in a black and white room, wears a black and white body suit and has never seen a coloured object. In other words, she has never *experienced* colour. This means that no matter how fully she understands the brain circuits that recognize colour, she will never have that subjective experience. This is the difference between the two aspects of consciousness. On the one hand, we have the brain and its wiring that allow consciousness to occur. On the other, we have some, presumably physical, process that generates subjective experience from those circuits.

Of course, this thought experiment can be criticized on the basis that black and white are colours and Mary experiences those. Also, as a human, she may be hard-wired to experience colour with whatever it is that gives us conscious experience. Nevertheless, it illustrates the difference between automatic circuitry and subjective experience.

It is not even clear why we need to be conscious. As David Chalmers of the Australian National University puts it:

Indeed, nobody knows why these physical processes are accompanied by conscious experience at all. Why is it that when our brains process light of a certain wavelength, we have an experience of deep purple? Why do we have any experience at all? Could not an unconscious automaton have performed the same tasks just as well? These are questions that we would like a theory of consciousness to answer.

This is a good example of the difficulty in understanding consciousness. David Chalmers uses philosophical *zombies* to illustrate his argument. By zombie, he means a being that looks and behaves like a human, but only in functional terms. The zombie has no conscious experience of the world, but responds to the environment and to itself and others as if it does. If questioned, it will respond as any other human would and will report having beliefs, such as knowledge that its car is in the driveway. The idea of a zombie world is logically consistent and is used by Chalmers to argue that there is an "easy" and a "hard" problem of consciousness. The easy problem is that of explaining function, but the hard problem is that of qualitative experience. An almost exactly oppo-site view is taken by the philosopher Daniel Dennett. He argues that if a being like this could exist, it would be conscious, and would experi-ence the world. We are all zombies. He takes a functional view and strongly argues against Descartes' idea of a separate, internal, observing conscious mind (a little person inside) that observes the world and acts. He proposes a "multiple drafts" model of consciousness in which com-ponents of the brain are continually editing and interpreting sensory data in parallel. This takes place over large fractions of a second and in different parts of the brain simultaneously, during which any special-ized part of the brain may edit or change information. We do not become conscious of these multiple drafts of edited sensory input at any particular time, but at any given time the "winning" narrative is the one we are aware of. We have a single idea of self despite these multiple narratives because we have an inbuilt system for self-definition which involves telling a narrative of who we are. Our "self" is a narrative spun out of these multiple narratives. The winning nar-rative is the one that is consciousness.

emergent properties and impossibility – taking advantage of design

If we take Chalmer's view, then the concept of a conscious being rather than an unconscious automaton performing tasks is a real mystery. Why should we be conscious? A view taken by Bernard J. Baars of the Neurosciences Institute in San Diego is that consciousness is a global workspace, acting as a blackboard on which the different subconscious processors can make their information available to the rest of the brain. Only informative and consistent information makes it to this workspace and therefore to consciousness.

The biologist D'Arcy Thompson pointed out that not everything in nature has been designed for its own sake. For example, if we want to build a room with a roof supported by arches, there must be spaces between their tops. This is a property of arches in a three-dimensional universe. The space between them has not been designed but is a con-sequence of supporting a roof that way. We may take advantage of it, hang a painting or colour the space, but the space was not put there for that purpose. We would not be able to build a roof supported by arches without these spaces – it would be physically impossible. It may be that consciousness is a similar thing. It may be impossible to build a brain with feedback circuits, inputs and outputs, designed to learn to recognize patterns and interact with the world, without subjective experience being generated. To try to do so without generating consciousness would be the impossibility. Because neural nets learn through interaction with the world, when we see a colour or an object or hear a word, it has a meaning. A computer doing the same thing would not ascribe a meaning, only accord a reaction. Meaning does not come from a web of definitions, but from interacting with the world and learning in a way that only a neural net can. By this definition, a computer that can learn from experience is different from a machine that is programmed to report its findings. It extracts meaning from the world and could be conscious. Recent advances in robotics are generating teams of football-playing robots that learn from their experience. These robots may be the precursors of truly conscious machines. Their descendants could be racist, sexist, lazy, generous, kind and have all the other traits we ascribe to conscious beings, because they would have electronic circuits that pattern-recognize and learn; designed to categorize the world and generate meaning from experience.

Many neuroscientists now believe that consciousness is an emergent property of the brain; a property that cannot be explained or predicted from knowledge of the parts. In other words, the system has properties that its components do not. A common example is crowd behaviour: a Mexican wave begins spontaneously, but knowing all the individual people making up the crowd would not help us predict the existence of a Mexican wave if we had never seen one before. It seems probable that, just like the crowd, neurons acting in groups somehow develop a property that is the neural basis of consciousness. One strong proponent of this idea is John Searle, of the University of California at Berkeley, who takes the view that consciousness cannot be reduced and is an emergent property. He argues that this means it cannot be considered objectively by understanding neurons. To do so misses the basic idea of consciousness as a subjective experience. Consciousness is a natural process of the brain in the same way as digestion is a natural process of the stomach, but the link between the nervous system and consciousness is a cause-and-effect relationship that does not rely on events. The release of digestive juices from the stomach results in digestion. The brain on the other hand has an effect like gravity, which is not an event, but causes effects like the pressure of an object on a surface. The brain equivalent is the generation of consciousness and this property emerges from the neurons and their connections.

Not all scientists believe this, though. The physicist Roger Penrose believes that consciousness has some properties that cannot be explained by large structures (neurons) obeying the laws of classical physics. Brain structures generating consciousness must instead obey the rules of quantum physics, because these are inherently unpredictable. This means the only structures that have the necessary properties to generate consciousness are *microtubules*, a part of the skeleton of neurons. We will spend a little time on Penrose's ideas, because they have had a great impact on the field and many major neuroscientists and artificial intelligence researchers disagree strongly with them. Penrose questions the fundamental dogma that most neuroscientists and lay people believe deeply – that neurons are the thinking components of the brain.

quantum physics and consciousness – Penrose's unreasonable idea

> The reasonable man adapts himself to the world; the unreasonable one persists in trying to adapt the world to himself. Therefore, all progress depends on the unreasonable man.
>
> George Bernard Shaw (dramatist and socialist, 1856–1950)

To appreciate why quantum physics might be involved in consciousness we need to learn about a mathematical idea, Gödel's theorem, and understand a little quantum theory. Although this might sound daunting, we need just a few of the ideas, rather than the proofs, so hopefully it won't be too heavy!

Gödel's incompleteness theorem – so we can't know everything after all

Gödel showed that it is impossible to completely describe the world using any rule-following system of symbols. This means that spoken or written language, mathematics, music, physics, computing and any other system will always be incomplete descriptions of the universe. There will always be insoluble and indescribable paradoxes and statements that will be impossible to prove, even if we know them to be true. Thus, any form of knowledge can never be complete.

The way Gödel proved this was to generate the mathematical equivalent of these two sentences:

1. The following sentence is true.
2. The previous sentence is false.

A moment's contemplation reveals that this is a paradox since, if we believe sentence 2 (as we are obliged to by sentence 1), then sentence 1 must be false and we are no longer obliged to believe sentence 2, which means sentence 1 was true in the first place, and so on. Gödel was able to show that such paradoxes exist in any system that uses rules, such as maths, computing or language. Humans use symbols internally to think about the world, for example in the form of language, so Gödel's idea should apply equally to us. This is where Roger Penrose's argument begins.

Penrose believes that Gödel's idea does not apply to conscious thought, because it only applies to systems that follow rules to manipulate symbols, such as mathematics or language, and our consciousness does not follow rules. He believes that mathematics could not have been worked out by any kind of rule-following problem-solving computer. If it could, we should be able to design a computer that could churn out all the mathematical proofs that exist and design new mathematical ideas as a result. To be able to generate original thought is beyond a rule-following system and, since we have original thoughts, we must be a system that does not obey rules and therefore our brains must be something more than classical computers. Consciousness is non-computational. Neurons and neural networks can be modelled on a computer or modelled with wires and circuits and connected together, but they are still subject to the laws of computing. Penrose's argument would mean that no matter how complex the network and how advanced the computer, it would never be conscious. It would always be prone to Gödel's theorem and could not, for example, come up with mathematics from scratch. There would be no way to make artificial intelligence. This also means that neurons and their connections are not responsible for thought and consciousness, because they could be completely and accurately modelled, given a sufficiently powerful computer. So why does Roger Penrose believe that microtubules are responsible for consciousness? To understand this, we need to know a bit about quantum theory.

quantum theory – microscopic madness

Early physicists thought that light was a wave. Waves cannot exist on their own; they need a medium to travel in: for example, sound waves travel in air, tidal waves travel in water. Light was said to travel in the ether, a substance that pervaded space and was also the medium through which gravity and other forces responsible for "action at a distance" could be transmitted. In a famous series of experiments, repeated over many years, the physicists Michelson and Morley showed that the ether did not exist. Einstein then discovered a beautiful set of laws, which formed the *theory of relativity* and which did not require that the ether existed. This theory had some counter-intuitive predictions. First, it predicted that there was an upper limit to how quickly things could travel; nothing could travel faster than light. It also predicted that time would pass at a different rate for someone travelling close to the speed of light. For example, one of a pair of twins who went up in a space-ship, travelled very fast and returned to earth after experiencing just a couple of weeks of spaceship time would find on their return that the other twin was very old, because, in Earth time, the spaceship would have been away for years. Unfortunately, the theory also meant that light could be a particle (it did not need a medium to travel in) or it could be a wave (it obeyed the inverse square law and had electromagnetic wave properties). The predictions of Einstein's theory have been tested on many occasions and, so far, our observations have been consistent with his predictions. Relativity explains how the universe behaves on a large scale, but it does not explain how atoms or subatomic particles behave. To explain the problem of whether light is a particle or a wave, other physicists came up with the *quantum theory*.

The cornerstone of quantum theory is that the only thing that matters is what can be measured. If a tree falls in a forest and no one is there to hear it, does it make a sound? In quantum physics, the answer is not straightforward. If something cannot be measured then we cannot know anything about it. If we can measure it, then the only truth we know is what we have measured. Unless the sound of the tree falling was recorded in some way, in a memory or on a machine, it cannot be said to have been made. In addition, the certainty we have about our measurement is reality. For example, we cannot be certain that a vacuum really is empty space, because the emptier it is, the more diffi-cult it becomes to be certain that a single atom has not escaped our notice. For quantum physics, since measurement is reality, a vacuum is full of virtual particles, coming into existence just briefly enough that we cannot be sure they exist and then vanishing again, just before we would notice them.

The first major consequence of this idea is *Heisenberg's uncertainty principle*: the problem that, for some pairs of properties, being more certain of one makes us less certain of the other, because we can only measure one of the pair at a time and the measurement destroys the other property. For example, we cannot know both the position and the velocity of an electron at the same time. To measure the velocity, we have to bounce particles off the electron to "see" it, but this moves it, so we no longer know where it was when we measured it. On the other hand, to find its position, we need to make it collide with something, but this destroys information about how fast it was moving.

The second major consequence is that, for some things, how we measure their properties affects what the properties are. We find what we look for. In this model of the world, light is a particle and a wave at the same time. How it behaves depends on how it is measured. If we look for a wave, it behaves like a wave. If we look for a particle, it behaves like a particle.

Physicists resolve these issues by saying that multiple possibilities are all true simultaneously. If a particle could be in one of several possible positions at any one time, then it is actually in all the positions simultaneously, until it is measured, but its probability of being in any particular place affects how likely we are to find it there when we measure it. If something could be a wave or a particle, then it is actually both until it is measured. Things exist as probabilities, becoming certain only as they are measured. Observation of something by a conscious being

SCHRÖDINGER'S CAT

In 1935, the physicist Erwin Schrödinger proposed a famous thought experiment, now known as *Schrödinger's Cat*. Quantum physics says that objects exist in all possible states simultaneously until observed by a conscious observer. For example, an electron is in every possible position around an atomic nucleus simultaneously (superposition of states), but once it is observed, its properties become real and fixed (collapse of the wave function). Schrödinger accepted that quantum physical rules applied to small particles, such as single atoms, but he could not accept that this was also true of classical objects, made of many atoms. To show how ridiculous the idea was, his thought experiment described a cat, locked in a lead box containing a radioactive atom linked to a poison vial. The poison is released, and the cat dies, only if the atom decays. The laws of quantum physics state that we cannot know if the atom has decayed or not until we open the box and check the cat. This means that until we open the box, the atom exists in both states simultaneously. Because the fate of the cat is directly linked to the fate of the atom, the cat is both alive and dead at the same time – clearly a ridiculous conclusion. Many of the crazy predictions made about quantum physics have proved correct and, in 2000, scientists from the State University of New York showed that for a large collection of atoms a Schrödinger's-Cat-type state could exist.

A TEST FOR CONSCIOUSNESS?

In Schrödinger's experiment, the cat could be regarded as a conscious observer, which therefore collapses the quantum field and resolves the paradox. We could design a thought experiment based on this idea which was a test for consciousness. A system could be set up and observed by the test being (that is, the cat). The result could then be observed by a second conscious being (us). If the experiment were devised correctly, it might be possible to make one outcome happen if the first observation resulted in collapse of the quantum state and a different outcome if the second did. That way, we would be able to tell if the first observer were truly conscious.

Modern versions of quantum theory see all energy as information and all matter as a way of computing information. Simple measurement of any kind (regardless of whether the measurer is conscious or not) is sufficient to collapse the quantum field, so it seems this test for consciousness would not work.

collapses its state from one of uncertainty into one of definiteness. Just as for relativity, the crazy predictions quantum physics makes about the world have all been found to be true – so far.

The problem is that in some cases (so far untestable), quantum theory and relativity make different predictions about the same things, for example gravity, so it is likely that one or the other will need to be modified to take this into account. It is in the quantum theory of gravity that Penrose has his basis for consciousness. Because quantum physics deals in probabilities, not certainties, a quantum particle, before it collapses into a definite state, has the non-computer-like properties of consciousness. Quantum theory only describes small particles, so neurons are too large to be the basis of a quantum theory of consciousness. Microtubules have several properties that make them a possible source for quantum effects to be harnessed. In them, a spherical protein, tubulin, is stacked into a circular arrangement and built up into long hollow tubes. These microtubules extend the length of neurons and vastly outnumber them. Because it does not quite stack neatly into the tube-like structure, the tubulin protein can have many different possible arrangements in any one microtubule. This means in quantum terms that it exists in all the possible states simultaneously. As the number of simultaneous states grows, the differences between mass and energy in the states reach a gravitationally significant level (so it becomes easy to measure) and this causes them to collapse to a definite state. This process is the direct cause of a conscious thought such as the decision to drink some water or wear black trousers. Penrose believes that this is a form of computing, but without the properties of a classical computer and therefore is not only beyond Gödel's theorem but also possesses the properties needed for consciousness. As further evidence of microtubules being the source of consciousness, Penrose argues that single cell organisms obviously have no nervous system, but make decisions and appear to learn: they contain microtubules.

Roger Penrose's arguments are fascinating and daring, particularly because he is a physicist, not a neuroscientist, and has therefore put up

IMPLICIT ASSUMPTIONS ABOUT CONSCIOUSNESS

When we think about what it means to be conscious, we often make several assumptions. For example, we may assume that consciousness must take place on a familiar timescale. We have no evidence for this. If we disregard this, trees could be conscious, because although they have no nerves, they do have a system that conducts information around the organism (xylem and phloem). These could function like neurons in very slow motion. Another assumption is that a particular size and complexity of nervous system is required. While we might think we have evidence for this, we only really do for nervous systems laid out like our own. How complex does a mollusc nervous system need to be? The Humboldt squid is able to flash in complex patterns and colours and is undoubtedly communicating when doing so. Its behaviour alters from peaceful and calm to highly aggressive and cannibalistic when it is fished by a method known as "jigging". This involves a barbed device that, once swallowed, cannot be regurgitated. The squid are dragged up violently by jerking the barb in the gullet, decompressing as they surface and with their entire body weight transmitted through the barbs. They are usually attacked by other Humboldts and decapitated or eaten alive. If we assume a simple flashing coloured mollusc, this seems bizarre behaviour from a gentle animal. If we assume a complex, conscious, communicating being, this behaviour might be a response by other Humboldts to a "please finish me off quickly" request from a captured squid in pain and distress. Our current thinking about consciousness does not allow us to consider that an animal like the Humboldt squid might be capable of such complex communication, or be conscious in the way that we are.

the backs of many people for "talking about things that are outside his area". It is difficult for most neuroscientists to believe that neurons are not the basis of consciousness, but, on the other hand, someone with Alzheimer's disease retains many neurons, but their microtubules collapse, so even a neuroscientist has to accept that things are not straightforward. In the end, it comes down to whether we really are more than highly complex neural networks and this is unlikely to be resolved with certainty in the near future.

the truth about consciousness

So, consciousness is either a non-existent problem, a problem that can never be solved, a problem that will be solved with computers that mimic neurons, or a property of microtubules, depending on whom we believe. In any case, we can marvel that we are able to think about it.

memory – putting the past in perspective

Why is it that our memory is good enough to retain the least triviality that happens to us, and yet not good enough to recollect how often we have told it to the same person?

Francois de La Rochefoucauld (author and moralist, 1613–1680)

Nothing fixes a thing so intensely in the memory as the wish to forget it.

Michel de Montaigne (essayist, 1533–1592)

types of memory – the long and the short of it

Just as sensation feels like a single experience, but is composed of different types of feeling, memory is not a single entity but a complex system of processes, all related, able to be separated by scientific enquiry.

sensory memory

The simplest division of memory is into *short* and *long term*. Most neuroscientists also recognize a third type, *sensory memory*, where all information received by the senses is stored before being passed to the conscious mind and stored in short-term memory. Sensory memory can store as much information as is received at any one time, but only for about a third of a second. If we do not attend to it, such information does not pass to the short-term memory.

short-term memory

Short-term memory lasts only seconds and is a temporary store before information is consolidated into long-term memory. Short-term memory holds about seven different pieces of information for about twenty seconds (although it can hold on to information for minutes at a time). Chunks of information count as a single piece, regardless of how complex each piece is, provided they can be coded as a concept. This means a hyphenated telephone number is easier to remember than a long string of single digits. Neurologists test short-term memory by asking people to repeat a series of random digits forwards and backwards. Most people can remember about seven digits forwards and four backwards.

Short-term memory consists of three types of memory store: *iconic*, which stores visual information; *acoustic*, which stores sounds; and *working*, which acts as a store for anything we need to remember to function (for example, repeating a phone number to ourselves or performing mental arithmetic, which requires that the digits are remembered while we do the sums).

Short-term memories are stored in the *prefrontal cortex*. When someone suffers an injury to this part of the brain or a tumour infiltrates it, they develop a permanent failure in short-term memory. We can mimic this in the laboratory by discharging a magnetic device over the front of the skull, which temporarily depolarizes the neurons in the prefrontal cortex.

Someone with a short-term memory failure can still use their long-term memory. Unfortunately, because short-term memory is used to organize what is needed for the situation at hand, they may appear forgetful. This is because they are poor at attending to the task of remembering or retrieving the information from long-term memory.

long-term memory

Transferring information from short-term to long-term memory is complex. It doesn't matter how long something has been in short-term memory, transferring the information requires either a level of emotional or intellectual understanding or a reduction in complexity. This process of transfer is learning and requires the conversion of information from electrical charge to biochemical and physical changes.

Learned information is useless unless it can be recalled and this is the other requirement of long-term memory. It is often thought of as having two different components; *explicit memory*, requiring conscious recall, and *implicit memory*, which is unconscious but affects our responses or abilities.

Consciously recalled information (explicit memory) consists of memories of personal experiences and lists of facts (*semantic memory*).

Memory of personal experiences is sometimes called *episodic memory*, because we remember the episode. For example, one day at school we learnt that London was the capital of the United Kingdom. Remembering what we did that day and that we learnt that London is the capital, is recall of an episodic memory, because it is a consciously recalled personal experience. Failure of episodic memory is one of the main features of Alzheimer's disease. If now asked what the capital of the UK is, we wouldn't necessarily recall the experience of learning it, but we would know that London is the capital, as a fact. This is a semantic memory.

Being able to ride a bike is an example of an implicit memory. We do not recall the exact bicycle-riding procedure consciously; we just know how to do it. Another type of implicit memory is a strong conditioned response, such as feeling excited by the prospect of going to the cinema. The previous memories of going to the cinema are not individually recalled, but they do have an effect on our response to the present.

memory circuits – time and emotion

In 1957, a medical report described a man whose severe epilepsy had proven impossible to control. As a last-ditch surgical treatment his hippocampus and amygdala were removed from both sides. He lost all his episodic long-term memory. A second report, twenty-seven years later, showed that there was no recovery of this memory loss. As a result of this, and similar reports, we now know that memory requires two important neural circuits. The first acts as a basic memory template and involves the hippocampus, its output wiring (the fornix), the front part of the thalamus and part of the limbic cortex. This system integrates the time component of a memory, so we have a concept of when something happened and how long ago that was compared with the present. The second system integrates emotion into memory and involves the amygdala, the top and middle of the thalamus and the prefrontal cortex. Without this, it becomes impossible to remember the emotional significance of an event and, because of it, we can remember how we felt at the time of a particular personal experience.

the memory address book

The role of the hippocampus is similar to an address book or postal code system and works for all types of long-term memory, although semantic (factual) memory can also use other circuits. The address is used as a tag to encode a memory and allows recall at a later date by returning to the same address. Addresses are not attached randomly to new information; rather, memories are incorporated into existing knowledge. In general, in a right-handed person, the left hippocampus deals with verbal memories and the right with visuospatial and musical memories.

The structures making up these circuits are very complex and so are vulnerable to damage. An example of this is the amnesia seen after a cardiac arrest, the result of damage to the large neurons of the hippocampus, because of the low levels of oxygen reaching the brain during the arrest. This causes *anterograde amnesia*, the failure to lay down new memories, although memories from before the damage are retained. If more of the hippocampus and other medial temporal lobe structures are damaged, the severity of the episodic memory disturbance is correspondingly worse. There is greater loss of past memories, though the most distant are better retained than more recent memories: the *temporal gradient* of memory. If damage occurs to the fornix – the major output wiring of the hippocampus – then the loss is permanent, even if the rest of the temporal lobe is intact.

the memory timeline – down memory lane

Damage to the frontal lobes causes a problem with the organization of memories. Because this part of the brain is also involved with planning, attention and concentration, it becomes difficult for someone with damaged frontal lobes to lay down new memories, but also the relationship of memories and time becomes jumbled. The person may start a conversation sounding as if they believe their grandfather is still alive but towards the end discuss attending their father's funeral and mention that he died of old age; not noticing the logical problems this involves. Along with the loss of temporal order comes a loss of the sense of personal memory or "knowing what one knows". The listener needs considerable patience, as the person will have no idea that their memory of the events is wrong or indeed that there is anything wrong with their memory at all.

confabulation – making it up as we go along

The most spectacular failure of the frontal lobe or frontal connections is *confabulation* (commonly seen in those whose memory failure is a result of alcohol abuse). Someone with this condition will make up clearly false stories or responses to questions, which they report as true. The causes are probably a mixture of failure in episodic memory, tem-

MEMORY FAILURE AND ALCOHOL

Alcohol has many effects on memory, both short term, such as the memory loss associated with a previous night of heavy drinking, and also long term. One severe problem is *Wernicke's encephalopathy*, in which extensive and excessive drinking leads to a deficiency of thiamine (vitamin B1) in the diet. This can lead to catastrophic bleeding into the mamillary bodies in the limbic system once the person eats, because of a surge in glucose in the blood stream. This can also occur because of other conditions that cause nutritional problems, such as prolonged morning sickness. The person may be uncoordinated and have eye movement problems. Even with treatment, up to a fifth of sufferers may die. Those who survive develop *Korsakoff syndrome*, a condition marked by severe short- and long-term memory problems. The person is unable to make new memories and has great difficulty recalling events from several years before the illness. They spend their time not knowing when or where they are, although they are otherwise alert and aware. They may also confabulate, as the brain tries to fill in the gaps that should not exist.

poral sequencing of memories, monitoring what is being said and comparing it to episodic memory (which itself is faulty). This makes it easy to mix up fragments of memories with imagination. Confabulation is particularly seen when the medial frontal lobes are damaged or disconnected.

remembering facts and figures

Semantic (fact-based) memory uses the same circuits as episodic memory, so that there is a time and emotion context for facts. It is also stored in other circuits, which means that semantic memories can be deeply embedded in our knowledge matrix. Once a semantic memory becomes embedded, it can be recalled without the emotional circuit, thus differing from episodic memory. Semantic memories are distributed throughout the cortex and are categorized – for example, the outer part of the left temporal lobe is the major site for storing nouns and names of living things. Within each storage region, certain categories of memories may be more vulnerable to damage than others. For example, if the speech area of the brain is damaged, a bilingual person may lose their second language but not their first. Even for monoglots, the names

of people they work with may be affected more than the names of
people they play sport with. Very specific categories can be affected,
such as losing the names of four legged animals but retaining the names
of two legged animals – thus being unable to name a lion but able to
name a kangaroo.

skill recall

The parts of the brain most involved with the implicit memory of skills
learning are the basal ganglia and the cerebellum. The skills themselves
are not accessed consciously, although the desire to perform them is. In
fact, attending to the action can worsen the recall, such as a golfer who
is thinking too much about how they should play a particular shot,
instead of "just doing it", or a musician who cannot remember how to
play a piece if they think about it. Interestingly, skills may be stored as
short sequences of memory and it can be difficult to access them from
the middle, as when someone who has learnt a piece of music by heart
is only able to play it from the beginning of a section.

examples of memory failure

Every so often, the headlines report a dramatic example of memory
failure, such as the surgeon who, in the middle of a heart bypass operation, suddenly forgot what he was doing, where he was and why he was
doing it. Fortunately, the extremely experienced senior theatre nurse
was able to tell him what to do next until the operation was complete.
Similar events happen reasonably frequently and are remarkable
because the affected person retains all their implicit memory, so they
remain skilled, and their previous episodic memories, so they know
who they are. What they lose is the ability to store or recall new episodic
memories, leaving them confused, anxious and desperate for answers
to the same repeated questions: "where am I?", "why am I here?" Attacks
of *transient global amnesia* may follow an electrical disturbance in the
brain, of the type seen in epilepsy, or possibly as a result of a stroke.

More commonly, memory problems creep up on us. The commonest cause of an apparently impaired memory is the failure to take proper
care in memorizing. This may seem trivial but is a cause of considerable distress and is especially common in older people. It was first
studied in the early 1990s by Jennifer Day at Sheffield University. She
asked two groups of people to tackle a memory test. The first group
were aged eighteen to twenty-five and the second were over sixty-five
years old; all had similar educational backgrounds. They were asked to

memorize the positions of eighteen objects placed on a 10×10 grid. Once a subject was sure they had looked for long enough, the objects were removed and they were asked to try to replace them in exactly the same positions. Just as you would expect, the young group did much better than the old group. The critical point, however, was that the subjects could themselves decide when they had looked for long enough to memorize the objects. The young group spent, on average, one and a half minutes (and in some cases nearly two and a half minutes) to memorize the positions, while the older group spent, on average, 42 seconds and in some cases as little as 25 seconds. When everyone was forced to spend the same length of time memorizing the positions, then the abilities of the two groups were almost identical.

Depression is another common cause of apparent memory impairment. Here, there is usually a more sudden deterioration than would be the case in a dementing illness, which fluctuates in severity and is usually associated with considerable concern about the memory loss. Someone with dementia will progressively worsen and often try to hide their failings.

The commonest cause of true memory loss is dementia and the most common type is Alzheimer's disease, in which there is damage to the limbic system and areas of the frontal, parietal and temporal lobes. There is an insidious loss of episodic memory, which is pronounced for recent events, whereas distant memories are preserved. Working memory is not badly affected, so that someone with Alzheimer's disease can remain independent, provided they can remember to remember. In tandem with the memory problems come changes in personality and behaviour and problems with orientation, in both time and place. As the disease progresses, the extent of the problems becomes wider and their severity worsens, until, eventually, memory fails for both recent and distant events, speech is empty of content with loss of comprehension and behaviour becomes more annoying with an exaggeration of previous personality traits, apathy and rigidity of personality. The change in behaviour and personality can be very difficult, with accusations laid about a spouse's fidelity, carer's honesty or apparent burglary requiring police intervention. As Alzheimer's disease progresses, the affected person becomes mute, unmoving and unable to recognize even close family members.

treatment of memory loss – a brainy pill?

The neurotransmitter *acetylcholine* is known to be important in memory circuits. If a drug which blocks the effect of acetylcholine is given to healthy young volunteers, short-term and episodic memory are impaired, although semantic, working and implicit memory are unaffected. In other words, it becomes difficult to learn new things or recall experiences, but knowledge and skills are unaffected. Working memory is still intact, so it is possible to repeat things or function from moment to moment. Because this pattern of impairment is similar to that seen in Alzheimer's disease, drugs that can increase the levels of acetylcholine in the brain have been used to treat its symptoms, with some success.

Working memory depends on the frontal lobes, which have a large number of receptors for the neurotransmitter *dopamine*. Given drugs that mimic dopamine, people who have a poor working memory improve, while those who have an excellent working memory deteriorate, which implies that there may be an optimal level of dopamine.

There is now ample evidence that greater use of the brain protects against dementia, but while training for a specific task does protect against its loss, it does not have a more general protective effect on the brain and memory function.

memories

To a large extent, our sense of self depends on memory. The personality can remain largely unaffected, even in people with damaged short-term memory, but our awareness of who we are, what we are doing and where we are going is intimately tied up with our ability to learn new information and recall the old.

chapter ten

sleep – bedtime for the brain

Sleep; King of all the Gods and of all mortals,
Harken now, prithee, to my word;
And if ever before thou didst listen, obey me now,
And I will ever be grateful to thee all my days.

(Homer, *Iliad*)

the history of sleep – at one with the universe

Sleep has long intrigued humankind and different cultures have tried to understand the healing, refreshing nature of sleep and its association with the altered consciousness of dreaming. The *Upanishad*, an ancient Indian philosophy text, considers human existence to have four states, all related to sleep: the waking self, the dreaming self, the self in a deep, dreamless sleep and the "very" self (a sort of super-consciousness). For the Chinese, sleep was regarded as a state of unity with the universe. In the words of the philosopher Chuang-Tzu in about 300 BCE: "Everything is one; during sleep the soul undistracted, is absorbed into unity." The mystical nature of sleep was thought to be related to healing and so plants that could induce it were presumed useful, not just because they could induce sleep but also because of its direct healing nature.

from the ancient Greeks to recent times

About 900 BCE, Homer described a chieftain, Asclepius, who came to be seen as a god of healing. People were brought to his temples in the hope that he might visit them in their sleep and cure them. It took about another five hundred years for a rational school of philosophy to develop and for people to start to think more logically about sleep.

PLANTS FOR INSOMNIA

Ancient medicinal plants included the opium poppy, belladonna and nightshade. Opium is derived from the juice of the unripe seed pods of the opium poppy. It takes part of its scientific name *Papaver somniferum*, from Latin for "sleep" and has probably been used for the treatment of insomnia since at least Sumerian times, about six thousand years ago.

Alcmæon, a Greek medical writer and philosopher-scientist, living around 450 BCE, was one of the first. He proposed that sleep was caused by blood flowing away from the surface of the body into large vessels and that we awake when it flows back into the body again. His ideas seem to have been taken up and modified by Hippocrates and Aristotle. Hippocrates developed a theory of the benefit or harm of sleep by observing its medical effects. He noticed that both excessive sleepiness and insomnia were undesirable and that people who were ill either slept a lot or were tired. On the other hand, sleep could restore the ill to health. Aristotle thought that imagination was the result of the senses perceiving an object even in its absence, meaning that the human mind can form after-images of things. In our waking state, we can distinguish real from imagined things, but this ability is lost in sleep, which is why our dreams can contain such fantasies. Aristotle therefore proposed that dreams are the product of the experiences we have while awake. In the seventeenth century, sleep was thought to be the result of animal spirits revealing themselves and, in the eighteenth, to come from poor circulation to the nerves. In the nineteenth century, it was suggested that sleep was due to reduced oxygen supply to the brain. None of these ideas was based on careful scientific observation and, as a result, none has stood the test of time.

The first truly scientific observations began in the early 1900s. Sleep-promoting substances were discovered in the spinal fluid of animals and soon afterwards it became possible to measure brainwaves directly by electroencephalogram (EEG). This provided a new insight into the function of the brain and opened a door to the understanding of one of life's greatest mysteries.

the neuronal origin of sleep – the sleep circuit

Current thinking is that we have specific neural circuits that keep us awake and if these are switched off we fall asleep. These neurons are in the *reticular activating system* in the brainstem. Signals from this system feed into the thalamus, which combines them with the sensory information it is receiving and relays it all to the cortex. The system uses a neurotransmitter, *glutamate*, which tends to activate nerves and therefore acts like a gate, allowing the passage of sensations to the thalamus and thence to the cortex. If this gate closes, we become insensible to the outside world, which is why we can sleep through noise or movement. There is a second wakefulness system in the hypothalamus, which is part of the autonomic nervous system and therefore responsible for regulating heart rate, breathing, sweating and other automatic processes. Signals spread from here along the base of the brain into the cortex. This system uses *histamine*, which is why antihistamine anti-allergy drugs block it and cause drowsiness.

These two systems keep us awake. Sleep-promoting neurons and chemicals produced by our bodies inhibit them, so that we normally proceed slowly towards sleep, taking about twenty munites. If we are sleep deprived, this process can speed up to just a few minutes.

When we are awake, our neurons fire in an organized, yet unpredictable, way. An EEG reflects this, showing apparently random, small, fast, noisy-looking waves. As we fall asleep, the neurons in the cortex begin to synchronize, so that rather than apparently chaotic firing, groups of neurons fire simultaneously and large waves of electrical activity appear. These *slow waves* become more and more prominent as we fall more and more deeply asleep. This process is governed by the thalamus, which blocks signals from the outside world, helped by the neurotransmitter *serotonin*, and locks in the cortical cells in a repeating loop. Serotonin also reduces movement, making it easier to relax. Drugs that block the effects of serotonin, such as Angel Dust (PCP), prevent sleep. On the other hand, if sleep is disrupted by depression, then both can be treated by increasing serotonin levels. Because of the synchronized nature of the neurons in this type of sleep, it is also called *synchronized sleep* and is arbitrarily divided into four stages, representing deeper and deeper states, where stage one is light drowsiness and stage four the deepest slow wave sleep. In total, we spend about eighty per cent of our sleep time in one of these four stages, with twenty per cent in stage four.

to sleep . . .

About an hour and a half after first falling asleep, something strange happens: the EEG becomes indistinguishable from an awake EEG and, at the same time, our eyes begin to move rapidly under their lids. This is rapid eye movement (REM) sleep (some people also refer to synchronized sleep as non-REM or NREM sleep). There are regular rhythmic fluctuations between REM and NREM sleep, every sixty to ninety minutes through the night, so that the total cycle happens about four or five times. Most NREM sleep happens in the first third of the night and most REM sleep happens in the early hours, before we wake. The first REM period is only about ten minutes, while the last is about an hour long.

Where does REM sleep come from? REM sleep originates in brainstem structures and has two basic components, the (so-called) *tonic* and the *phasic*. In most people these occur simultaneously, but they can occur separately. Tonic signals paralyse the muscles (except for the diaphragm and eyes) and switch the cortex into an awake pattern; phasic signals travel into the thalamus and are relayed on to the cortex. Every burst of phasic signals is followed by a burst of rapid eye movement. Other phasic bursts cause occasional muscle twitches and activate the autonomic nervous system, causing increased breathing, changes in heart rate, pupil dilation, penile erection and decreased sweating.

. . . perchance to dream

Awakened from REM sleep, eighty-five per cent of people will report a dream. Most will report a visual experience and around sixty-five per cent will also report auditory experiences, such as music. Spatial experiences, such as falling, flying or floating are rarer, while taste, smell or pain perceptions are very rare. Functional blood flow scans suggest that the parts of the brain associated with complex motor activities such as walking, running and swinging are also activated during dreams but cannot be expressed, because of the motor paralysis, preventing unintentional injury. People or animals with a disturbance in the tonic pathway, who thus cannot paralyse their muscles, act out their dreams. It is not clear whether the rapid eye movements of REM sleep are observations of visual dream experiences but people blind from birth have them, which suggests they are not.

It is thought that the phasic signals arriving at the cortex are interpreted by the awake, but disconnected, cortex as dreams, to which, it is speculated, we give meaning, depending on our past experiences and current anxieties. It isn't clear whether the dreaming experience is random or organized. We know that bilingual people dream in the language appropriate to the rest of the dream setting; for example, in English when dreaming of England and in German when dreaming of experiences set in Germany or with German-speaking relatives. This suggests that there is some system to the dreaming experience. There is also an emotional component of information processing, which allows memories to associate more freely than during the normal waking state. Thus dreams may not only be highly emotionally charged, but can also be attempts at emotional problem solving.

Dreaming is a natural mental activity. Most dreams are clear, coherent, realistic and detailed accounts of a situation involving the dreamer and other people. More often than not, dreams are about very ordinary activities and preoccupations, although they can also be of fantastic or ridiculous things. Dream content is influenced by the sleeping place – dreams at home are more emotional and personal, with a greater amount of sexual or aggressive content than those in a laboratory. Dreaming occurs in both REM and NREM sleep, though the content in NREM sleep does not generally follow a simple narrative, as it does in REM sleep. It tends to become more like REM sleep dreaming if the subject becomes anxious or unhappy.

Dreaming can be looked upon as a continuous stream of mental activity, which we become aware of when aroused, occurring particularly during REM and to a lesser extent during NREM sleep. Dreaming is a difficult area for research, because of the difficulty in separating the effects of waking consciousness from the consciousness of dreaming.

. . . ay, there's the rub

Dreams can affect waking and personal boundaries between fantasy and reality, and influence their content. People with clear boundaries may be able to switch off bad dreams better than those with more fluid boundaries, such as creative people or those with mental illnesses like schizophrenia.

sleep architecture

New-borns sleep in multiple bursts through a twenty-four hour period. As the day–night rhythm becomes stronger, these bursts generally become consolidated into two periods of sleep, one at night and one in the afternoon, a pattern established by four or five years of age. Over the next few years, this consolidates further until, by the age of ten, most children have a single overnight period of sleep, although in some cultures the pattern of two periods of sleep persists into adulthood, for example with an afternoon siesta.

Whereas an adult spends about a fifth of sleep in REM sleep, a new-born spends about a half. This gradually reduces, over the first six months, to become more like the adult pattern. A new-born can fall directly into REM sleep, rather than first passing through NREM sleep, but this ability is lost by about the first year. As we age, we tend to spend less time in the deeper stages of sleep and have a more fragmentary sleep pattern. This natural change is regarded as frustrating by many, who resort to sleeping tablets.

The human is not the only animal that sleeps. All mammals show REM and NREM sleep, although some marine mammals sleep with each half of their brain in turn. Certain branches of the mammal line, such as the Australian echidna, do not have distinct REM and NREM sleep. In contrast, birds have separate REM and NREM states (although, again, some birds can sleep with each half of their brain in turn). Insects and spiders have two different patterns of sleep, but these are not identifiable as equivalent to REM and NREM. Because sleep is a phenomenon seen throughout the animal kingdom, it must have an important function. Despite its clear importance and our increasing understanding of how it happens, we do not know the underlying reasons for sleep. This has lead to a number of theories, some of which hark back to ancient times and the Greek philosophical ideas we discussed earlier.

why do we sleep?

Popularly, sleep is for restoration, a theory supported by evidence that hormones that increase protein production are released during sleep, while hormones that increase protein consumption are switched off. The longer we go without sleep, the stronger the need for NREM sleep, so some people think that NREM is to restore the body and REM might be to restore the brain.

Another idea is that sleep is a way to conserve energy. Animals with higher metabolic rates sleep longer. There is a reduction in brain metabolism and blood flow during NREM sleep, but this effect is lost in REM sleep, when metabolism and blood flow return to waking levels. The average person "saves" just 120 calories during eight hours sleep, so this is not a very convincing reason.

More recently, the importance of sleep in memory and learning, and to the integrity of synaptic and neural networks, has become evident. If we try to learn a new task when deprived of REM sleep we do considerably worse than if deprived of NREM sleep, suggesting that we need REM sleep to lay down new memories. In contrast, we perform an already learnt task more poorly when deprived of NREM sleep, suggesting that recall or motor skills might be affected.

A further idea is that sleep is a way of stimulating neglected nerves, making sure they stay in tiptop condition. If muscles are left unexercised, they waste away; stimulating nerves during sleep might be a way of keeping the brain in good health. It has been suggested that, because REM sleep is associated with paralysis, it may be important for motor circuits, which can be stimulated without harm, while NREM sleep may be more useful for maintaining non-motor circuits.

sleep deprivation

Surveys in America suggest that, over the last one hundred years, there has been a reduction of one and a half hours in the average duration of a night's sleep. While the extent of this change within different populations may be difficult to assess, it does raise questions about the effects of sleep deprivation. Up to one-third of young adults suffer from excessive daytime sleepiness, as shown by a standardized scale (such as the Epworth sleepiness scale). In middle age, at least seven per cent of people suffer from excessive daytime sleepiness related to sleep disorders, while another two per cent are affected by shift work. This may not seem a great number, but the consequences can be catastrophic, when accidents such as the Chernobyl nuclear plant disaster, the Three Mile Island nuclear accident, the Exxon Valdez oil spill and the Selby train crash occur as a result of fatigue. More recent studies have shown an association between medical misadventures and sleep deprivation. Most road traffic accidents involving no other cars occur during the two periods of the day when humans are particularly vulnerable to sleepiness: from 3 to 5am and from 3 to 5pm.

If sleep deprivation is associated with accidents, can we decide how essential sleep is? Indefinitely depriving someone of all sleep is not ethical but is recognized as an effective technique in enforced questioning and classed as torture by the UN. It is likely that indefinite sleep deprivation would result in death. More limited periods of sleep deprivation have been investigated in volunteers, with a maximum recorded period of sleep loss of eleven days; the effects are both behavioural and physiological. The greatest behavioural effect is increased sleepiness. Intellectual performance is maintained for up to fifty hours, if a suitably stimulating diversion can be found, but the quality of the performance starts to decline after thirty-six hours. Subjects become irritable, have difficulty concentrating and can become disoriented. They may suffer visual illusions and hallucinations and may become paranoid and psychotic. Impairment of memory soon becomes a major feature. Physiological changes, apart from a minor decrease in body temperature, are low. Partial sleep deprivation, where total sleep time is restricted to five hours a day for some weeks, has also led to impaired performance and altered mood. Fortunately, all the effects of sleep deprivation can be treated by allowing unlimited sleep.

sleep disorders

Most people jerk a few times in the transition from wakefulness to sleep. These *hypnagogic jerks* are quite normal and have been likened to the final spark of a light bulb when turned off. There are, however, many types of experience or behaviour associated with sleep or partial arousal that are not normal, or are normal but quite distressing. These are known as *parasomnias* and frequently begin in childhood.

The first category of parasomnia is of problems occurring in the sleep–wake transition, such as meaningless rhythmic movements or behaviours such as body rocking, head banging, head rolling and body shuttling. They typically occur at the start of sleep, in the transition from wakefulness to early NREM sleep, though they can also be seen later during sleep and may be confused with epileptic seizures.

The second category is NREM sleep arousal disorders: sleep drunkenness, sleep-walking and sleep terrors. In these conditions, there is a partial arousal from deep sleep; the person may appear to be awake but fails to respond normally to commands. Often the person is distressed, but may be very resistant to consolation and may even become violent, with no recollection the next morning. Sleep-walking occurs in the first third of the night's sleep; the person may be clumsy and injure themselves. Sleep terrors usually occur suddenly; the person comes straight out of deep sleep, unlike a normal awakening, and is confused. These are associated with fear, a piercing scream, wide open eyes and a pounding heart. These attacks are frequently followed by uncontrolled running.

Rapid eye movement sleep disorders form the third category of parasomnias. These include nightmares, sleep paralysis and REM sleep behaviour disorder. In a nightmare the person wakes fully oriented but aware of an experience in which they feel as though their life was in danger. In contrast to sleep terrors, these events occur in the final third of the night's sleep, the person rarely rushes out of bed but needs consoling and is unable to go back to sleep for some time because of anxiety. Sleep paralysis usually occurs at the onset of sleep or immediately on awakening. The person remains fully aware but is completely unable to move, because the normal muscle paralysing system of REM sleep has not been switched off. It is extremely frightening and, although it only lasts at most several minutes, may feel like it has lasted for hours. Rapid eye movement sleep disorders often occur in normal individuals and may run in families. They may point to a more serious disorder such as narcolepsy, a condition in which there is an uncontrollable urge to sleep and no NREM on falling asleep. This means the person dreams immediately, which can make it difficult to distinguish dreams from reality. Chronic sleep deprivation can mimic some of the features of narcolepsy. Finally, REM sleep behaviour disorder occurs where the dreams of REM sleep are acted out because the paralysis normally present during REM sleep has failed. In adults, it is usually the sleeping partner who is most aware of these attacks and they may have to sleep in another bed.

the mystery of sleep

Sleep is still largely a mystery, but with scientific study and the instruments we now have, we are gradually unlocking its secrets.

the motor system – making movement and motion

I learned to walk as a baby and I haven't had a lesson since.
Marilyn Monroe (actress, 1926–1962)

skilled in movement

The motor system is the intimately connected collection of nerves and muscles that allows us to move ourselves and bits of our bodies. Human beings have the most advanced motor system on the planet. We can walk, run, swim, climb and crawl. We can balance on a single toe and spin, we can perform somersaults and hand springs, we can control a fast moving ball with absolute precision and we can write. The motor system allows us to perform our normal daily tasks, of course, but even these are incredibly sophisticated. Talking to another person requires the co-ordination of the larynx and vocal cords with the diaphragm, lips and tongue, at the same time controlling facial expression and most of the rest of the body with gestures, posture and movements appropriate to what is being said. Writing requires precise control of a tool that depends on pressure and speed to work effectively. Driving requires manipulation of a fast-moving complex machine while interacting with others.

dependent on feedback

All these tasks depend heavily on feedback from our eyes, balance organs and joints. Our eyes, for example, can remain fixed on a moving target while our heads are being jolted around and the rest of the body is moving too. This requires an unbelievable amount of control. The system that performs this tracking is different from the system we use to deliberately look at something. You can easily demonstrate this with a friend: ask them to try to pursue an imaginary object with their eyes and you will soon find that they can only make instant, jerky movements of their eyes, unless there is something real to follow, such as a moving finger, in which case the movement is a smooth pursuit. These jerks or *saccades* are what we use to look from one object to another and while our eyes flit between targets we are blind for a microsecond. Smooth pursuit movements, on the other hand, require a moving target to fix on and are dependent on the apparent movement of the target and our personal movement, as determined from the balance organs. So, our eyes have sophisticated movement systems, designed to provide us with the maximum information as rapidly as possible, allowing us to move around our environment gracefully. We do not need our eyes to do this, however, and we are graceful even in the dark, because we can feel where our limbs are in space – the sense of *proprioception*. This is combined with information from stretch receptors in muscles, so that we know how much tension there is in a limb, and with information from our balance organs, so we know where we are in space.

moving in harmony

How does the motor system work? The full answer to this question is not yet known, but we do have many parts of the answer. The motor system can be thought of as three separate systems, all of which are needed to allow us to move. We will explore them in no particular order, since no one part can be said to be more important than any other. Perhaps the most logical to start with, since it is the simplest conceptually, is the voluntary motor system.

the lead vocalist

When we decide to make a movement, the planning begins in the very front of the brain, in the *premotor cortex*. The precise relationship between planning and carrying out the movement is not known, but the first part of a voluntary movement seems to start in the *motor cortex*, the region of the brain just in front of the fissure between the parietal and frontal lobes. When we concentrate on or are conscious of a movement, this is the part of the motor system we are using. The nerve cells here are arranged in six layers and the largest cells are pyramid shaped.

It is these pyramidal cells that seem to start off the process of move-ment. Their axons are very long and travel in a large bundle, crossing, in the medulla, from left to right and vice versa. The bundle then con-tinues into the spine. The axon of the pyramidal cell enters the grey matter of the spinal cord and joins, at a synapse, with the cell body of a second nerve cell in the frontmost part. The axon of the second nerve cell then travels out of the spinal cord, bundled with other axons, down the limb to a synapse with a muscle fibre. The first neuron, from the brain to the spine, is the *upper motor neuron* and the second, from the spine to the muscle, is the *lower motor neuron*. Humans are unique among "higher" animals in having a part of the motor system that con-sists of just two nerves.

Each lower motor neuron is supplied by about ten upper motor neurons. Damage to the lower motor neurons causes a floppy, wasted muscle that may twitch, which may happen after suffering a broken limb. Damage to the upper motor neurons causes a stiff, weak muscle, often seen after a stroke. One disease causes death of upper and lower motor neurons, with no obvious reason and no known cure; it is relent-less and leads to progressive paralysis and death within a few years. This disease is motor neuron disease and only humans are known to be affected by such a condition. Other conditions, with similar names, can affect motor neurons in horses and dogs, but not in the same way. It may be that it is the simultaneous simplicity and sophistication of our motor system that makes it vulnerable.

the feedback monitors

There is a feedback loop, the *reflex arc*, from the muscle to the lower motor neuron. This is what makes us withdraw a hand from danger or prevents us overstretching a muscle without thinking. The reflex arc includes a sensory organ, the spindle (a modified muscle fibre, lying in among the other muscle fibres). When the spindle is stretched, it sends a signal back through the reflex arc, which makes the muscle contract strongly and quickly. No thought is required, since the signal does not reach the brain. It goes from the spindle to the lower motor neuron in the spinal cord and back out to the muscle. This means that muscle con-traction from a reflex trigger is much faster than a normal contraction, because the signal does not have far to travel nor many connections to jump. The spindles can be very relaxed or very sensitive and this adjusts the degree of tone in our muscles. There are similar fibres in the tendons that detect a dangerous level of stretch and switch off the muscle con-traction completely, which prevents a muscle or tendon from tearing if we try to contract it too suddenly or strongly. This too happens at a spinal level, so we do not need to think about it at all.

the orchestra

Making a voluntary movement is all well and good, but what happens to the muscles we are not thinking about? If someone throws a ball, they are concentrating on their throwing hand and arm, not necessarily thinking about their other arm, back, legs, head or trunk, yet all these muscles are needed for correct throwing. Also, they may be thinking about which muscles they want to contract to throw the ball, but not about which muscles to relax. Bending an arm requires us to tense the muscles that bend the arm and to relax the muscles that straighten the arm, but the arm needs to be a little rigid to support itself and so the relaxed muscles must not relax completely. In general, if a muscle that bends a limb is contracting, the opposite (antagonistic) muscle, which straightens the limb, is relaxing. This unconscious muscle control is taken care of by the basal ganglia, tucked away deep inside the brain on either side of the thalamus.

The basal ganglia control the background activity on which a conscious movement is made. They are like the backing music to the lead vocal, or the rest of the team to the football player with the ball. They are not where the action is, but unless they are working properly the action is not what it should be. When we are sitting having a cup of tea, we may be thinking about lifting the cup to our lips, but the rest of our body is in a position and state that is appropriate to drinking tea. If we stand up, our body needs to change the muscular background, so that it is now appropriate for standing, and then if we walk, it needs to change again into a walking mode. When the basal ganglia do not work properly, the muscular background becomes disordered. If they are underactive, it becomes difficult to change quickly from sitting to standing or from standing to walking. The state of relative relaxation and contraction of different muscle groups is also disordered, so that rather than antagonistic muscle pairs being relaxed or contracted appropriately, they are all contracted, leading to rigidity. This scenario of difficulty changing or starting actions, combined with excessive stiffness and often a tremor, is seen in Parkinson's disease, the classic disease of underactive basal ganglia. In the opposite state, when the basal ganglia are overactive, the muscular background is constantly changing or inappropriate to the voluntary movements being made. It may be impossible to keep still, which is what happens in *Huntington's chorea*, a disease in which a person becomes more and more fidgety and restless. This disease is also associated with dementia. Overactive basal ganglia may lead to inappropriate body positions for the voluntary movement being carried out, as happens in *writer's cramp*, when the hand holding the pen adopts an abnormal position and tension, making it impossible to write easily.

Although we have to learn certain actions to be able to make them consciously, once we can perform them, we do not have to think very hard about making them, even if they are very complicated. This is because of the third component of the motor system: the cerebellum.

the conductor

The cerebellum is the "little brain" sitting on the back of the brainstem, at the back of the head. It is very densely wired, with axons in a peculiar tree-like arrangement, which gives it another name, "the tree of life". Unlike most other parts of the nervous system, the nerves of the cerebellum control the same side of the body: the right cerebellum controls the right body. The cerebellum has many functions, but the most important is as a storage depot for movements we have learnt. Motor programs are stored here, so that once we know how to ride a bicycle, we do not need not learn again; we can just call up the bicycle-riding program and there is a template of basal ganglia and motor cortex activation that we can modify for the particular situation we find ourselves in. The more we practice, the better the stored motor program becomes and the more we can concentrate on fine tuning our performance by concentrating on one part of the action (or even on nothing), so that the action is entirely automatic. The cerebellum integrates the learned action with information from the balance and joint position senses. In other words, the cerebellum is responsible for co-ordination. Problems with the cerebellum, as may happen in multiple sclerosis, affect the co-ordination of actions. But the commonest problem affecting the cerebellum is intoxication: being drunk affects our ability to walk, talk or perform other co-ordinated actions. The cerebellum may even co-ordinate mental processes, so that some scientists believe it has a role in consciousness and thinking too.

balance – the neglected organs

The balance organs are specialized motion detectors, located in the base of the skull. This means they are really motion detectors of the skull rather than of the whole body but this is not a problem, since the rest of the body has joint position sense with which to compare the skull

THE WRITING ON THE WALL

If we write with a spray can on a wall, the writing looks the same as when we write with a pen on paper. This is quite surprising, because writing with a pen requires small movements of the wrist and fingers, whereas writing on the wall needs movements from the shoulder and elbow. If motor programs consisted of which muscles to relax and contract our writing would look different depending on the muscles being used and we would not be able to write on a wall without learning how to. This means that learnt movements are stored as concepts and activated motor programs trigger muscle contractions that will fulfil the concept that is required.

position. The balance organs consist of three *semicircular canals*, which are sensitive to rotation movements in the three dimensions of space, and two sack-like organs, the *utricle* and *saccula*, which are sensitive to head tilt and moving or stopping in a straight line. The utricle is sensitive to vertical movements, while the saccula is sensitive to horizontal movements. When we are flying, the semicircular canals detect banking, the utricle detects the climb, descent and turbulence and the saccula detects acceleration and deceleration. All five structures depend on the relative movement of fluid against fine hairs within hollow tubes or bags. The bending of the hairs in the fluid current, like seaweed moving in the sea, leads to the activation of an action potential in the balance nerve. This signal then makes its way to the cortex via the thalamus.

Information from the balance organs splits in two in the brainstem. Part of the signal goes to the region of the brain controlling eye movements, which allows us to fix on an object and keep tracking it, even if our head is moving and we are in a moving object. This system is incredibly fast, accurate and sophisticated and allows us to keep images absolutely fixed on a minute part of the retina. If there is a problem with the balance organs or the eye movement control, we will have the flicking eye movements of nystagmus. The other part of the signal goes to the regions of the brain controlling limb and trunk position, so that we do not fall over (unless we want to). This makes sure our weight always passes through our centre of gravity, which is always over our feet.

the performance

How are the three systems that make up the motor system connected? Signals from the motor cortex travel both to the basal ganglia and directly down to the spinal cord. The cerebellum has three input/output sections, the *cerebellar peduncles* or *stalks*. Motor cortex axons travelling down into the spinal cord meet with axons of the cerebellum coming out of the middle stalk in the pons and their signal is modified. Meanwhile, signals carrying information about where the body and limbs are in space are collected by the lower stalk, processed by the cerebellum and sent up to the thalamus. Signals from the motor cortex, processed by the basal ganglia, also go to the thalamus. The thalamus acts as a relay station for these sensory signals and sends them back, as feedback, to the motor cortex. The three parts of the motor system are intimately connected and orchestrate our actions, so that we only need to concentrate on our desire, while the actions needed to achieve it, both for the rest of the body and for the correct co-ordination of the part we are thinking about, happen automatically, smoothly and elegantly.

the sensory system
– feel the world

Pain (any pain – emotional, physical, mental) has a message. The information it has about our life can be remarkably specific, but it usually falls into one of two categories: "We would be more alive if we did more of this," and, "Life would be more lovely if we did less of that." Once we get the pain's message, and follow its advice, the pain goes away.

Peter McWilliams (*Life 101*)

It takes a wonderful brain and exquisite senses to produce a few stupid ideas.

George Santayana (philosopher and poet, 1863–1952)

Feeling is one of the five senses, as any school child can tell us. This simple word hides a complex and sophisticated armoury of sensory apparatus, fine-tuned to detect our environment in many ways. Most of these detectors are located in the largest organ of the human body, the skin and are designed to receive information about the outside world, but some are turned inwards, to detect our inner environment.

no sense no feeling

What is the sense of feeling? At least six different senses make up what seems to be a single sense. First is the sense of *pain*, important to all animals and probably present in every organism with a nervous system. Second is the sense of *temperature*: we are not particularly good at telling the difference between hot and cold and usually need other cues to know which way to perceive the sensation. Third is *light touch*, the sense we are probably most aware of most of the time. It is what we use to actively feel something, but we make the most use of it in combination with other feeling senses. These three senses map on to skin and are located in it. Under the microscope, a wide variety of feeling detectors can be seen and, at first, scientists thought that each detector could only detect one type of sensation, but we now know that a single receptor can detect different sensations. In other words, there is no specific detector for heat, touch or pain; rather, any detector can respond to heat and any to touch. Even so, some are better at detecting particular types of sensation, so there is a little bit of specialization going on.

The next sense, *vibration* sense, is best felt by bony parts of the body. The most vibration-sensitive parts of the body are the teeth and, of these, the canine teeth are the most sensitive. It may seem strange that this is a sense on its own, but the ability to feel vibration is a distinct entity and provides a sense of pressure.

The fifth sense gives us an awareness of where our limbs are in space. This is *joint position* sense or *proprioception*. The receptors for this are based in joint capsules and allow us to walk in the dark or to type without looking at the keys. They provide important feedback when we are judging distance, for example reaching for a cup of tea or bringing food to our mouth. This sense is intimately related to co-ordination and its nerve pathways are strongly connected to the cerebellum.

The sixth sense is a synthesis, performed at a high level in the nervous system. It is not a sense with its own nerve fibres and connections, but is generated by the brain when it receives all the information from the other senses. This is *two-point discrimination*, the ability to tell if an object in contact with the skin is a single point or more. On the finger tips, we can distinguish points only one or two millimetres apart. Any closer and they can feel like a single point (you can test this with a friend, your eyes closed and the ends of an unfolded paperclip). On the back of the hand, the points need to be much further apart to be distinguished and on the back of the trunk, it can be difficult to distinguish separate points that are centimetres apart.

strength of feeling

Although each sensation detector can respond to any stimulus, its connections determine what we experience. A detector connected to a pain nerve will make us feel pain, if stimulated strongly enough. The same detector connected to a temperature nerve will make us feel heat or cold. The intensity of the feeling depends on how fast the nerve is firing and on how many nerves from the same area are firing. Strong sensations are caused by rapid firing of many neighbouring nerves. Gradually, the detectors and nerves become used to the stimulus and the intensity of firing slows down until a change in the situation re-activates them. So, we have a system that can detect different sensations and signal how strong they are, but how can we tell which area is being affected? This depends on where they connect in the brain. Each part of the thalamus and the parietal lobes corresponds to a part of the body, so that a signal arriving in that region is felt as a sensation in the corresponding part of the body.

The types of sensation are in three groups. First are pain and temperature. Nerve fibres carrying this type of information travel straight to the thalamus, where they split into two. Some are relayed to the sensory cortex and are used to localize a sensation accurately. Others are relayed to the limbic system, which controls arousal and emotion. These are what make us sweat with pain or feel sick or angry. Second are vibration and joint position sense. These are fast nerve fibres, allowing us quickly to adjust our body position. They start their journey in the rearmost part of the spinal cord. Third is two-point discrimination, generated by the cortex.

There are other senses of feeling, some of which are purely reflex, such as the sense of muscle stretch and the sense of tendon tension, but the six we have mentioned are the ones most people are actively aware of each day. Of these, we will concentrate most on pain, because this is a good model for the other senses of feeling and we know quite a lot about it.

pain – can't live with it, can't live without it

Pain is an important sense; some would say the most important. Without pain, it is surprisingly easy to suffer injuries and lose bits of the body. This is why leprosy is thought by many people to lead to limbs dropping off, whereas it actually damages nerve fibres, so that the person remains unaware that digits or limbs have been injured, leaving them prone to infection or further injury. Similar things can happen following nerve damage as a result of diabetes, or in *syringomyelia*, a condition in which a cavity in the spinal cord specifically damages the nerve fibres dealing with pain and temperature sensation. It is vitally important to be able to experience pain, but when pain is extreme, chronic or unrelated to an obvious cause, it becomes less useful and more distressing. In these situations it would be helpful to be able to switch it off. This can be done, to a greater or lesser extent, by interfering with the pain pathway anywhere on the route from its source in the skin or body to its perception in the brain.

The things that can cause pain vary from one tissue to another. For skin, it is injuries like cutting, crushing, extreme heat or extreme cold. For hollow organs such as the stomach or gall bladder, it is induced by swelling in the walls, stretching or spasm of the muscle wall of the organ or traction on the ligaments holding the organ up. For ligaments, stretching or inflammation of neighbouring joint linings can cause pain, but, strangely, cutting a ligament does not. Stretching a blood vessel or compressing a nerve may also trigger pain. For muscles, pain is caused by being starved of oxygen, death of muscle fibres (as happens in some diseases), swelling of the muscle and prolonged contraction. Most of us have experienced this last cause of muscle pain when running a long way or holding heavy things for a long time. It happens because muscles can create energy from sugar in two ways. The standard way is with oxygen, but if oxygen is in short supply, or if the exercise load is so heavy that a little extra energy is needed, muscles can switch to anaerobic metabolism: a process similar to that yeast uses for fermentation. Human muscle, rather than creating alcohol as a by-product, as yeast would, creates lactic acid, which causes pain and stiffness. If evolution had given us just one different enzyme in this metabolic pathway, we would make alcohol when we exercised and it might be a lot more popular, particularly among young people!

changing events into feelings

How do all these different processes become converted into an electrical signal that can travel up a nerve and be perceived as pain? Most activate a system of hormones and chemicals in the injured region, in the process of *inflammation*. The main protagonists in this reaction are histamine, prostaglandins, serotonin, some small proteins and potassium. Any one of these can bind to pain receptors and set in motion the action potential that travels down the nerve and into the nervous system. Some of the small proteins make the pain receptors more sensitive, although they do not themselves cause pain.

Once the pain stimulus has been converted into an electrical signal, it travels down the pain nerve and into the spinal cord. Pain nerves are not insulated with myelin, so they conduct signals more slowly than other nerves. This is why pain can sometimes take a little time to make itself known. Because of the reflex loops in the spinal cord, we can withdraw from the pain source automatically while the pain signal is still on its way up to the brain, leaving us with the experience of reacting before we know why, as you will know if you have ever accidentally touched a hot oven or iron.

REFERRED PAIN

Localizing pain coming from internal organs can be difficult. We usually perceive deep internal pain signals based on embryological development. For example, the *appendix* is a small structure at the end of the small intestine, lying in the lower right corner of the abdomen. If we have an inflamed appendix, the pain is first felt around the belly button. This is because the appendix is in the mid-gut, which, in development, was at the level of the middle of the abdomen. When the appendix becomes further inflamed, it begins to stretch an internal tissue, the *peritoneum*. This makes it easier for the brain to localize the signals and the pain moves to the lower right of the abdomen. If the appendix then bursts, the inflammation spreads around the peritoneum until it reaches the under-surface of the diaphragm. Again, the brain cannot localize the pain this causes and rather than perceiving it as being under the diaphragm, we feel it in the shoulder tip. This is because the diaphragm starts off at shoulder level in the embryo and migrates down to the adult position during development.

Another classic example is heart attack pain, which may be felt in the centre of the chest and may radiate down the left arm. This is because the arms are part of the chest embryologically, and bud out as we develop. The radiating arm pain is really chest pain, but the chest is now an arm.

And if a particularly painful experience happens early in our childhood, later pain, from nearby structures, may be inaccurately perceived. For example, someone who suffered bad sinus pain as a child may feel toothache as a sinus rather than a tooth pain.

In the spinal cord, the pain nerves connect not only with the reflex loop but also with the chain of nerves climbing up towards the brain. The signal travels up the spinal cord and into the thalamus. All senses arrive, in some form, in the thalamus, which processes them in some way before sending them on to the cortex, to be integrated into our conscious experience. Because the thalamus has such close connections to the cerebral cortex it is not possible to separate clearly their roles but we now think the parietal lobes of the cerebral cortex are doing the main work of whatever it is that makes us consciously aware of sensation, although in the past, the thalamus was thought to be the seat of consciousness.

blocking pain at source

Experiments have shown that the threshold for pain is similar for most people. This means that the lowest intensity stimulus that can be recognized as pain is about the same for everyone. This is modified by personality and the rest of the brain, so, for example, people with disconnected frontal lobes react to pain either briefly or not at all. People with a neurotic, anxious personality have the same threshold for pain as non-neurotic individuals but react to it differently.

Inflammation lowers the threshold for pain (sensitization), so that a stimulus that would not normally be painful might become so. We can block the inflammation pathway with aspirin or paracetamol, which reduces sensitization and blocks some of the direct triggers for pain. A more effective method is to raise the pain threshold with local anaesthetics. These agents stop action potentials travelling down a nerve, so that the pain signal never reaches the spinal cord or the brain. This is not always practical, because local anaesthetics tend to stop all sensory nerves from working properly, not just those dealing with pain. Someone with a painful arm might not want a completely numb one, and in any case, the large doses needed for a whole arm would have other effects on the heart and the nervous system. Fortunately, we can intervene further up the pain pathway.

closing the pain gate

In the spinal cord, where the pain nerves connect with other nerves, is the *pain gate*. Nerves dealing with touch sensation can switch off the connections between a pain nerve and the rest of the chain, which means that a pain signal arriving in the spinal cord cannot go any further and the pain sensation never arrives in the brain. We have all experienced the pain gate in action by "rubbing it better". This contact with a painful area activates the touch nerves that block the pain nerves. This makes sense from an evolutionary point of view, because it is a signal that the organism is aware of the pain and is taking action. The chemicals that do the blocking are the *endorphins*, our body's natural form of morphine, which act on receptors on the nerve cell surface, the *opioid receptors*. This is one way in which opium derivatives like morphine or heroin reduce pain. Another method that can stimulate and close the pain gate is *transcutaneous electrical nerve stimulation*, or TENS. A TENS machine delivers a low electric current which activates and closes the pain gate, thus relieving pain. This can be highly effective for some types of pain, including the labour of childbirth.

disconnecting pain from experience

Once pain signals make it past the pain gate, the route is not completely clear. On the way to the thalamus, nerve fibres carrying pain signals pass through a region of the midbrain, the *periaqueductal grey*. This region is full of opioid receptors and has a strong pain-inhibiting effect. Some of the nerves in this region reach back down into the spinal cord and connect with the pain gate, helping to close it. Stimulation of opioid receptors by heroin or morphine has a disconnecting effect on pain. The pain is not decreased, as it might be with a local anaesthetic, but its unpleasant quality is removed and it can feel remote. People being treated for severe pain describe the sensation with statements like "I know there is a pain somewhere in the room, but I cannot be sure I have it."

There are other mechanisms that reduce pain. Placebos (medically neutral substances, like sugar pills) are effective painkillers in about a third of people, working through an, as yet, unknown mechanism. Acupuncture may stimulate endorphin release in the body, but this is just a guess and, in any case, why it should do so is not clear. Distraction techniques, such as hot or cold packs, or extreme emotional states such as fear, anger or mania may all act to reduce or delay the experience of pain. Perhaps the most extreme example of the innate drive to use distraction comes from the condition of *cluster headache*, a rare form of migraine that can affect men (usually but not exclusively, heavy drinkers). Although it is now treatable, the pain in this condition is so severe that it has driven people to suicide. More typically, it leads the person to bang their head hard and repeatedly against a wall, because the distraction of the head injury pain is preferable to the cluster headache pain. Another example comes from soldiers in combat. Studies have shown that soldiers wounded in battle initially require little or no pain relief but subsequently need more, when away from the front line and in hospital. The distraction and disconnection system is probably active whenever we are under stress or pressure.

sensation – why bother?

Why do we need to feel at all? Plants seem to be fine with just light and gravity information. The answer lies in the motor system. If we did not move, we probably would not need to feel in quite such detail. This is why the parts of our body with the most sensory nerve endings are those that most explore the environment: lips, tongue and hands. This is echoed by the amount of brain power for sensation perception allocated to each part of the body. The brain is organized so that between neighbouring specialized areas there is a region of overlap, where signals from the two are mapped together: the *association areas*. It's perhaps not surprising that the sensory strip, which lies just behind the fissure separating the frontal and parietal lobes, is next to the motor strip, which lies just in front, meaning that information about movement and sensation can be exchanged quickly. Some people believe consciousness comes into being at this sensorimotor interface.

the visuospatial system – making light work

seeing is believing

> Pictures, propagated by motion along the fibres of the optic nerves in the brain, are the cause of vision.
>
> Isaac Newton (natural philosopher, 1642–1727)

Newton was certainly on the right track and, though our current understanding of the sense of vision is far greater, there is much that still remains a mystery.

Perception is how an organism detects and interprets the external world. Humans do it in five ways – smell, taste, sound, touch and sight. The sense of sight comes at a high price, requiring the largest single proportion of total brain computing power; much of the cortex, either directly, or indirectly through supporting senses like proprioception or memory, is given over to visual processing.

From an evolutionary perspective, the development of this specialized cortical region, the visual cortex, is most significant. In his book *In the Blink of an Eye*, the zoologist Robert Parker argues that the development of the eye prompted an explosion in the number of different species of animal inhabiting the Earth during the Cambrian Period. The ability for predators to see prey, or prey to see predators, put enormous selection pressure on all species, so that the genetic mutations associated with small refinements in the visual system conferred survival advantage. The development of a visual sense raised the stakes hugely, particularly if you were an erstwhile simple creature, with just a few cells capable of responding, very slowly, to chemicals in the environment. Imagine for a minute that some humans developed the ability to read the thoughts of others. This might allow them to predict hostile intent and dodge an attack or buy shares at the right time and make enough money to retire early – either way, it would confer a potential survival advantage over those without that sense. In its impact on the course of evolution, the development of vision was as revolutionary as this science fiction.

blindness and the brain

To imply that vision is somehow the "greatest" of all the senses, representing the "pinnacle" of evolution, would imply, quite wrongly, that those without this sense somehow have a reduced quality of life. On the contrary, many blind people report a far greater environmental awareness, through the development of other senses, notably hearing, smell and touch. This is not to pander to the stereotype of the blind piano-tuner. The massive cortical power that would have been handed over to vision can be diverted and experiments suggest that this is most evident in those who are blind from birth, less so in those blinded later in life. This fits with other experiments that have led to the view that plasticity, the ability to reorganize brain function, is at its most efficient in the first two years of life. For someone who is blind, the sense of touch works at the "speed of sight". This is clear when reading Braille at speed; something few (or no) sighted people could achieve. Some experimental results suggest that the visual cortex is still being activated during

WHAT DO BLIND PEOPLE SEE?

Sighted people often think that a blind person sees what they would see when they close their eyes or that they see black. People who were never sighted see the same as we see from our finger tips – nothing. This is not the same as seeing black. People who have become blind see nothing or little, but they may experience hallucinations consisting of very bright (sometimes painfully bright) colours or complex forms consisting of patterns, scenes, animals or people combined with normal perceptions. These hallucinations are the result of a sensory-deprived visual cortex, desperate to interpret the few signals that may come through the neural net. The hallucinations are not usually unpleasant but the experience can be worrying for some people, because they think they are going mad.

this process: in other words, the area "programmed" for vision has been recruited to process other sensory information instead.

the structure of the eye

Essentially, all that is required for some sort of vision is a light-sensitive cell or two. What refines this process is the fine tuning and pattern detection, the ability to focus the light reaching the light-sensitive cells and then to interpret what is detected. The interpretation is the really clever bit, and, like most of neurology, we gain much of our knowledge of these processes from what happens when it goes wrong. The common analogy with a camera is useful but, by having two eyes and the ability to move them in a co-ordinated fashion, we have three-dimensional vision and greatly improved image quality and usefulness.

Light enters the eye through the pupil and is focussed by the lens, which sits just behind it. Lining the inside of the eyeball is a layer of photosensitive cells: the retina. Most of the retina is made up of long thin "rod" cells, which produce electrical impulses in response to light. The colour-sensitive "cone" cells are concentrated at the centre of the retina – the *macula* or fovea. The eye is designed so that the majority of the light from the central part of the image is focussed on to the macula. This is the part of the retina with the highest image resolution and contains a high density of detection cells. The peripheral rod cells, further out on the retinal surface, function better in low light, but the image they produce is not so sharply focussed. All the information about the image must travel beyond the retina to the visual cortex, at the back of the brain.

THE QUALITY OF THE IMAGE

If you were to put a camera film in the back of the eye, in place of the retina, the resulting picture would be so poor it wouldn't be worth developing. The reason we see so well is entirely down to the amazing image processing hard-wired into our visual pathway, most of it in the retina. This image processing is able to edge enhance, detect straight lines, contrast enhance, colour enhance, detect movement and allow for continuous movement. The image enhancement system is responsible for some optical illusions, such as the appearance of dark blobs at the intersections of criss-cross patterns of white on black and also causes the echoes of complementary colours we see if we have been looking at a single colour for some time.

THE (NOT SO) BLIND SPOT

If you look at the surface of the retina you will see the billions of nerve fibres from the rods, cones and image enhancement cells all coming together in one large bundle, the *optic nerve*. This is positioned just off the midline of the retina. At this point (the optic disc or blind spot) there are no light detection cells and so no light is perceived. Why then do we not have two large holes in our visual fields? Perhaps each eye compensates for the hole in the vision of the other – no, close one eye: there is still no obvious hole. This is because the brain "fills it in" or neglects to make us aware of it. You can demonstrate your blind spot using a playing card. Take the Two of Hearts and hold it, horizontally, a short distance away. Look at the inner heart with the other eye closed. Keep looking at the inner heart, move the playing card slowly away and you will notice the outer heart disappear as the light from it falls on the blind spot.

Examination of the optic disc can give clues to the pressure inside the closed box formed by the skull and serve as an early warning of problems.

the visual pathway – the information superhighway

The vast amount of information from the retina travels to the occipital cortex to be processed. The visual "map" of the scene we are looking at is maintained during this transfer, so, for example, the words we are reading are perceived in the order they appear on the page rather than at random. Just as the entire skin surface is mapped on the surface of the brain, so the retinal surface is mapped in the visual cortex and information from different parts of the retina travels in different nerve fibre pathways.

visual fields – slicing up the retinal cake

Having two eyes (binocular vision) is the great strength of the human visual system, not least because it permits the detection of depth. However, the visual "map" must be divided between the two eyes to be transported to the brain, yet the equivalent parts of each retina must travel together, which requires the mixing, and subsequent division, of the pathways. This happens just behind the eyes where the two optic nerves come together, at the *optic chiasm*. From this point, the left halves of each retinal output, which represent the right halves of what we see, go to the left side of the brain and the two right halves (the left sides of the image) travel off down the right side. To complicate things just a little further, the retina is also divided along its horizontal axis, so that information from the top half (representing the bottom of the actual image) travels through the parietal lobe and the bottom half through the temporal lobe, on their way to the visual cortex.

Damage along the course of this wiring system gives rise to characteristic patterns of loss in the visual field. This enables neurologists to localize problems. For example, a stroke commonly damages one half of the brain. This will result in the affected person missing all (or part) of one half of their visual field, but this will be the same part of the field in both eyes. The person might fail to notice objects in this part of their vision, for example bumping into walls on one side.

the visual cortex – where images are interpreted

The occipital cortex is where the visual data arriving from the optic nerves are processed. Within this visual cortex there are specialized circuits responsive to size, movement and colour – all damage can result in impairments in the perception of any of them independently. There are up to twenty different parts of the visual cortex, but most are poorly understood. Viewed under the microscope, the neurons in the main part of the visual cortex (termed V1) appear to be in striped layers; these cells seem to respond particularly to elongated objects and edges. Each neuron in the visual cortex corresponds to a region of the retina. Some neurons respond to edges or shapes of a specific size, while others are less fussy. Some require the shape to be moving in a specific direction as well as being oriented a particular way. Neurons of region V2 are the most enigmatic, but may be responsible for detecting angles between pairs of lines. Those of V3 seem to respond to orientation, colour and depth, but not motion. Layer V4 has neurons that respond to colour and spatial information while those of layer V5 (also called MT) respond to motion. Each layer receives inputs from other layers so that the whole is a complex feedback system.

blindsight – seeing but not perceiving

By far the most fascinating of the syndromes affecting the visual cortex is *cortical blindness*, which can arise through a variety of diseases that damage the occipital cortex. One form of cortical blindness is blind-

PALINOPSIA – VISION STUCK ON PAUSE

Another interesting visual cortex phenomenon that can occur as a result of damage from strokes and sometimes migraine is *palinopsia*. In this condition, an image persists after it can no longer be seen. It is as if you were looking at the Sydney Harbour Bridge, then turned away to face the Opera House instead but the image of the bridge persisted.

sight, in which people claim to be unable to see, yet, if challenged, can perform visual tasks, such as posting a letter through a slot. The signals from the eye reach the cortex, but cannot be interpreted. Another form of cortical blindness was seen in a badminton player who complained that, although he could play perfectly well while the shuttlecock was in the air, once it came to rest on the ground he could no longer see it. Here, the conscious perception of movement was intact, but not that for still objects.

Blindsight has been used in films and television programmes to send subliminal messages to normally sighted people, by showing brief images too fast to consciously see, although they are subconsciously perceived by the brain, working at much higher speeds. This has been used for advertising and also in the film *The Exorcist*, where a single frame, showing a skull, was used to add subliminal fear. It is thought to take around half a second to perceive that something has been seen, but, since chemical transmission between nerve cells can occur in around one millisecond, thousands, even millions, of nerve connections may have been made before one is actually "aware". Many countries ban the use of these subliminal messaging methods. A similar phenomenon occurs with sound. We have all experienced being woken from sleep by a noise that seems to have occurred just after we actually woke up. This is another example of our conscious perception lagging behind the brain's processing of the information. Interestingly, the bright flash that often accompanies such a loud noise is thought to be "leakage" of nerve activity from the temporal lobe auditory cortex into the occipital lobe, where it is interpreted as a flash. This is the phenomenon of *synæsthesia*.

synæsthesia – mixed up in the senses

Synæsthesia refers to the mixing of senses, where the stimulation of one sense causes the perception of another, rather like crossed telephone wires. It is harmless and thought to occur in about one in every two thousand people, although recent studies suggest it may, in mild forms, affect as many as ten times this number. The commonest form is for people with synæsthesia to report that certain sounds, smells or tastes are associated with vivid, highly specific, colours. For example, they might perceive the number four as orange, or the telephone ringing as red. Gottfried Leibniz (1646–1716), the philosopher and mathematician, and Isaac Newton (1642–1727), scientist and mathematician, were both pioneers of calculus, but are also credited with references to the concept of synæsthesia. Leibniz recounted the tale of a blind man who perceived the colour scarlet as the sound of a trumpet and Newton noted parallels between the colours of the spectrum and musical notes. The painter David Hockney (b. 1937) has arranged lighting and set designs based on his synæstheric perception of the musical score.

the parietal lobes – integration of the supporting senses

Vision is more than just the reconstruction of an image. As well as seeing something, the brain requires information about meaning and about the body's relationship to objects in the image. The parietal lobes are critical for spatial processing; how we interpret our three-dimensional bodies and the world we live in. They are the great "integrators" of information from different parts of the brain, bringing visual information together with aspects of memory (via the temporal lobes) and sensorimotor cortical information.

There is a vast amount of sensory information arriving from all over the body at any one time, not just from the five major senses but also from lesser-known ones, such as joint position sense. Joint position sense is greatly strengthened by associated visual information. You can test this yourself: try standing on one leg, first with your eyes open and then with them closed. The task is much harder with eyes closed, since the brain is made entirely reliant on information from the joint position cells in the ankle joint and from the vestibular system of the inner ear to make the small corrections needed to keep you upright. Many people who sprain their ankles refer to their "bad ankle" forever afterwards, because they have also damaged the ankle joint position sensors. While their eyes are open, in the normal course of events, they have no problems, but once they start running, especially on uneven ground, the visual system alone is unable to keep the ankle in exactly the right place and they tend to keep re-spraining it.

If one parietal lobe is damaged it can result in *neglect* of the opposite side of the body, which might manifest itself as lack of self-care on the affected side – for example when dressing and washing. People with such neglect syndromes have been known to eat from only one side of their plates (which can be overcome by teaching them to remember to turn the plate after they think they have finished). This is not a problem with seeing the whole plate. Although visual pathways do pass through this region on their way to the visual cortex and, if damaged, can cause loss of vision over one half of each eye, this does not produce symptoms of neglect. The problem is more complex: altered appreciation of the three-dimensional world. One side of it ceases to exist conceptually. Another striking example is the clock face drawn by a person with neglect due to parietal lobe damage – all the numbers are present but crammed into one half of the clock face.

People with parietal lobe damage may also have difficulty in concentrating on more than one object at a time or may not be able to recognize, by touch alone, an object placed in the palm of the hand. A similar condition is the inability to recognize letters or numbers drawn on the palm of the hand. In the parietal lobes, we see further evidence of lateralization of certain functions. Damage to the dominant parietal lobe can result in *Gerstmann's Syndrome* (Josef Gerstmann, Austrian neurologist, 1887–1969), which includes right–left confusion, impairment of writing, impairment in arithmetic and the inability to distinguish the fingers of the hand. Damage to the non-dominant parietal lobe can cause difficulty in making things, but with a subsequent denial of these deficits. Finally, the inability to recognize a face, *prosopagnosia*, can also result from damage to the non-dominant parietal lobe.

While these syndromes are rarely as pure and as simplistic as described, they do provide an indication of the complex nature of the parietal lobes, which remain one of the most fascinating and possibly least understood of all the brain regions.

language, hearing and music – making sense of sound

> If the English language made any sense, a catastrophe would be an apostrophe with fur.
>
> Doug Larson (US newspaper columnist)

language – the world in symbols

Living in groups where strong social interaction occurs requires a certain amount of communication. In primates this occurs through body posture, eye gaze, physical contact and position. Grooming, which requires use of the hands, is an important part of social interaction and primates spend a lot of time doing it. One primate, which uses its hands for grazing, has taken the step of vocal grooming. These animals sit together in large herds and make continual noises, sounding like speech. If, in our evolutionary past, we required prolonged use of our hands, we might have become very vocal in the same way. Similarly, chimpanzees use verbal and hand signals to hunt in teams.

Mentally, we have the ability to encode the world around us as internal symbols; this, our internal dialogue is, in terms of time spent, the main use of language. The second use is to communicate these symbols to each other. In one sense, we each have our own unique language. The meaning we ascribe to symbols is personal and we can only understand each other because one person's symbols map to another's compatible set. If this fails for sufficient numbers of symbols, we say we are talking a different language.

Until recently, it was thought that any animal and therefore any brain could potentially learn a language. The ideas of Professor Noam Chomsky, of Harvard University, changed all that. He put forward the proposal, radical at the time, that language is a skill unique to humans. Just as a bird is specialized for flight, we are specialized for language. This proposal is at the heart of the philosophy of language. Are we hard-wired to speak, or would our flexible brain make the necessary connections anyway? The answer is probably a bit of both. We have specialized areas of the brain with the right connections and wiring tendencies to enable them to pick up a language, but we are not specifically designed to learn any particular language. Any baby growing up in a Japanese-speaking environment will learn flawless Japanese – not just Japanese babies. There is plasticity in the system. Our brain is designed to acquire language automatically, but it can acquire any language.

Phonemes, sounds such as "oo", "ma", etc. are the elements of language. A child without writing skills has the ability to move the tongue, lips and palate so as to be able to pronounce any phoneme, although stringing them together may be more difficult. Learning to write seems to reinforce the phonemes a child is using but causes the loss of the ability to make the unused phonemes. This is why adults learning a new language will always have an accent, whereas a truly bilingual individual, speaking both languages from childhood, will not. It is as if having a symbol for a phoneme gives it a permanent existence in the brain.

brain areas dealing with language – talking the talk

The auditory cortex is in and around the Sylvian fissure of the temporal lobes, and consists of several specialized regions. The comprehension of speech is carried out by a dedicated part of the brain, *Wernicke's area*. In almost all right-handed people, this is in the left hemisphere, at the meeting point of the temporal, parietal and occipital lobes. This region also holds the *angular gyrus*, where the internal voice is thought to be generated. Damage to Wernicke's area (usually caused by a stroke or tumour) results in "receptive dysphasia" – the inability to understand language: everything sounds like a foreign tongue. People with this damage speak fluently, but because the comprehension area is damaged, the words spoken are meaningless. This is described very well by the medical jargon "word salad"; randomly spoken and made-up words and phrases, said with the conviction of someone speaking sense.

Speech output is controlled by *Broca's area*, a region just in front of Wernicke's area, at the meeting point of the frontal and temporal lobes.

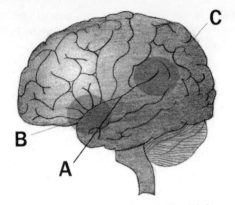

Figure 12 The auditory cortex: Broca's and Wernicke's areas. A: Sylvian fissure, B: Broca's area, C: Wernicke's area. Broca's and Wernicke's areas follow the general brain rule of: "behind the midline = input, in front of the midline = output".

Broca's area is in charge of fluency. Damage here leaves a person able to understand everything but unable to get the words out to reply: "expressive dysphasia". The feeling we all have occasionally, when a word is on the tip of our tongue but we cannot quite remember it, is the feeling someone with a Broca's area problem has all the time. This leaves people frustrated. Usually some speech is left, perhaps single words or phrases, such as "thank you" or "yes", but they may not be said appropriately, because although the person knows what they want to say, the connection between the internal desire and the verbal symbol is broken: the effort of saying something means that anything comes out. At best, sentences will consist of short, broken blocks of words with none of the usual conjunctions: "and", "to", "as", etc. If the level of speech output is a little higher, word-finding difficulty may lead someone to describe an object instead of naming it: for example, they may say "the thing you eat with" instead of "spoon".

For people who are multilingual, language disturbances usually affect the most recently acquired languages first, whatever the type of damage. We are not made so that one region outputs English and another outputs Spanish. If that were true, then it would be possible to lose either language, depending on where the damage was. We seem to be wired so that there is a brain region for language in general, but wiring for the first language is more robust to damage than wiring for later ones.

EVOLUTION OF A LANGUAGE

Recently, scientists obtained a rare insight into how languages develop when studying a group of deaf children in a school in Nicaragua, who, in isolation, had invented their own sign language spontaneously. They taught it to new students, who made it more sophisticated within one generation.

To describe to another person that a ball rolled down a hill quickly, but without using words, most people will make a rolling action with their dominant hand and imitate a ball rolling down a hill quickly. This is what the inventors of the language did. The next generation of children who grew up with the language modified it, so that instead of a single action denoting the entire concept, there was a symbol for the ball, another for rolling, another for the direction, and another for the speed. This is the same structure as a spoken language and is much more flexible. This shows that a simple language can quickly become complex and the concepts within a language can arise within a single generation.

Although we have said that humans seem to be specialized for language, writing is a relatively recent development, probably only a few thousand years old. This is nothing from an evolutionary point of view, which means that writing is either an inevitable consequence of spoken language or that it can be achieved with the brain specializations we already have.

Although we have talked a little about the parts of the brain dealing with language, we have not discussed a fundamental aspect of spoken language: the ability to hear.

hearing – air and waves

To understand hearing, we must first understand what sound is. A sound is a series of compressions and rarefactions of a substance such as air or water, with the wave travelling away from the source. The closer together the crests of the waves, the higher the pitch of the sound we hear. Humans can hear sounds as low as 20 Hz (twenty bunchings up a second), and as high as 20,000 Hz. But how is a series of compressions in air converted to the perception we call sound?

LANGUAGE SHAPES HUMAN THOUGHT

In 2004, a counting study of a Brazilian tribe supported the idea that we can only easily think of concepts for which we have words; in other words without a language we cannot think. This *Linguistic Determinism* was first proposed in the 1950s, but has remained controversial. Previous experiments have shown that while babies (and animals such as rats, pigeons and monkeys) can precisely count small quantities, they can only distinguish approximately between clusters of larger numbers. One possible explanation is an inability to articulate numbers. Hunter-gatherers from the Piraha tribe have no words for numbers above two, describing all such numbers as "many". A team of US scientists from Columbia University, led by Peter Gordon, carried out a series of experiments exploring the Pirahas' ability to handle the concepts of four, five or more. In the simplest experiment, a random number of familiar objects were laid out, and the person had to respond by laying out the same number of objects in their own pile. The Piraha could consistently match the numbers for one, two or three objects, but for four or more objects could match only approximately and their performance worsened the larger the number. In other experiments, they could not correctly recall if a box seen seconds earlier had four or five fish on the lid and could not mimic strings of four or five taps accurately, although they could for three or fewer.

This is evidence that a language that lacks certain words actually prevents speakers from being able to understand the concepts that would be embodied by those words and shows why specialist professions develop their own jargon.

Sound waves arrive at the *eardrum* and vibrate it, which, in turn, moves a series of small bones, the *malleus*, *incus* and *stapes*. The final bone, the stapes, is connected to an inner eardrum, the *oval windows*. The lever effect of these three bones means the sound arriving at the eardrum is amplified about twenty-two times by the time it enters the inner ear.

How are these amplified vibrations converted into the electrical signals our brain needs? The answer lies within the magic of the inner ear – a truly remarkable organ – the only completely mechanical organ

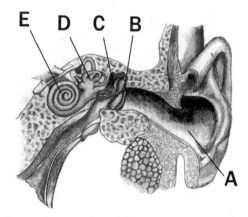

Figure 13   The ear. A: external canal (the "earhole"), B: tympanic membrane or eardrum, C: bones of the middle ear, D: balance organs – the semicircular canals, utricle and saccule, E: cochlea. D and E make up the inner ear.

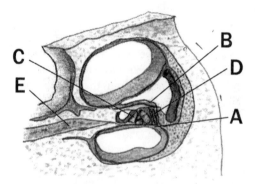

Figure 14   A section through the cochlea to show the inside of the "snail shell". A: vestibular membrane, B: basilar membrane, C: tectorial membrane, D: organ of Corti, E: auditory nerve. This is the "business end" of the cochlea.

in the body, it converts mechanical energy directly into electricity, in a beautiful and elegant way.

the ear – a hairy snail

The cavity of the snail-shell shaped inner ear is split in three along its length by two sheets of tissue. The lower of these sheets, the *cochlear*

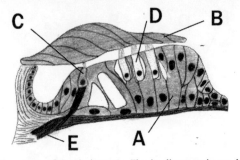

Figure 15 The organ of Corti close up. The basilar membrane is designed to move in response to sound waves. The inner hair cells have their "hairs" floating in the fluid below the tectorial membrane. The outer hair cells fine-tune the system. A: basilar membrane, B: tectorial membrane, C: inner hair cell, D: outer hair cells, E: nerve fibres becoming the auditory (also known as vestibulocochlear) nerve.

partition, is a two-ply sheet, the bottom ply being the *basilar membrane* and the upper ply the *tectorial membrane*. This structure converts vibrations into electricity. Because the basilar membrane is narrow and stiff at the bottom end and broad and flexible at the other, vibrations lead to a wave that travels up the sheet, starting at the narrow, stiff end and travelling towards the broad flexible end. The wave becomes bigger and more peaked as it goes, peaking at a point that depends on the frequency (pitch) of the wave. High frequencies peak near the narrow end, low frequencies at the broad end.

Hair cells are the key to the whole of our hearing ability and their remarkable arrangement is what gives them the power to detect minute sounds and yet respond up to twenty thousand times a second.

The hair cells have a bunch of cilia coming out of the top. The cells are embedded in the basilar membrane, with the hairs floating in the fluid above. They are arranged with the longest in the centre and the shortest on the outside. Each hair is connected at its tip to its longest neighbour by a very thin filament and "hinged" at its base. This means that any movement tilts some long hairs away from attached short hairs. Because the long hair will bend more, this creates a tension in the connecting filament as it pulls on the shortest hair cell. This system can detect a movement less than the width of an atom, corresponding to a sound on the very threshold of hearing. Each hair cell is connected to a nerve fibre and sends its electrical signal into the brain stem. While a simple sound will stimulate one hair cell the most, a complex sound stimulates several simultaneously. The mechanics of the system mean

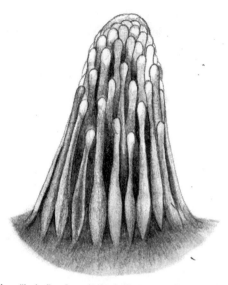

Figure 16 The "hairs" of a hair cell are really modified cilia called stereocilia.

that much of the signal processing happens before the signals even reach the nerves, let alone the brain.

The hearing system is remarkable for two more properties. The basilar membrane does indeed respond to sound as expected for a sheet with its properties, but it is able to distinguish notes that are almost identical in pitch at a resolution far higher than should be possible. There are only 3,500 inner hair cells: although an artificial basilar membrane with this many detectors might analyse sound effectively, it would not be as accurate as the human ear. How then are we able to pick out sounds and pitches so accurately? For every row of listening, inner hair cells there are three rows of "outer" hair cells that work backwards. Instead of listening, they are talking. The brain sends signals to the outer hair cells, making them move and therefore make a sound. It is thought this helps to fine-tune the basilar membrane and also acts as an amplifier for specific sounds (so that we can pick out a conversation in a noisy room – the "cocktail party effect"). This brain output going to the ear can be so strong that it is sometimes possible to hear the ear making a noise. These *oto-acoustic emissions* are the basis of a hearing test in babies. A tone is made and a microphone listens for the echo from the ear. If it is heard, this means that the hearing system in the baby is working.

The other remarkable property is the ability to detect sound direction. In owls, bats and dolphins, this has reached extremely sophisticated levels, but even in humans it is well developed. With two ears, it should be possible to tell if a sound has come from the left or right by its volume. This is not the only clue however. The sound arrives at one ear a few millionths of a second before it arrives at the other; this gives the strongest information about direction. There are specific groups of nerve cells in the brain that only respond to sound coming from particular directions. The shape (and position – most people have one ear slightly lower than the other) of the ear also allows us to determine if sound is coming from above, below or behind using similar clues, even though this will not affect the relative volume of the sound reaching each ear. With such a sophisticated hearing system, we can do more than listen to speech or noise from the general environment. We can also hear a third type of sound: music.

music and the brain – a symphony outside produces a symphony within

> Music . . . can name the unnamable and communicate the unknowable.
> Leonard Bernstein (composer and conductor, 1918–1990)

Modern humans have been playing bone flutes, percussion instruments and jaw harps for at least thirty thousand years. Neanderthal humans have left bone flutes that are over fifty-three thousand years old. All the human societies ever known have had music and musical appreciation seems to be hard-wired into us. Pre-lingual infants as young as eight weeks old respond to musical sounds, but turn away from unpleasant discordant sounds. Music has definite emotional effects on us and brain scans show that it lights up the same pleasure areas of the brain as do sex, addictive drugs and chocolate. But why does music exist? Either musical appreciation has a survival advantage and is therefore selected by evolution, or it is the accidental result of having a brain wired as it is for other reasons (like D'Arcy Thompson's arches – see chapter 8). Studies have shown that there is probably no specialized brain centre for music; it is analysed by many different parts simultaneously, including regions that are used for other things, such as vision or speech. The exact areas used vary depending on the experiences and training of the listener and are very adaptable, revealing great plasticity, so that even a small amount of study can change the way music is dealt with by the brain.

music and speech

Historically, neurological conditions leading to musical problems have been used to understand how music is processed. Two composers with such difficulties led researchers to the idea that speech and music are handled separately by the brain. In 1933, Maurice Ravel developed a condition in which he became specifically unable to write music. He could still hear music and could play scales and his speech and language abilities were unaffected. This could be thought of as the musical equivalent of the speech output problem seen in damage to Broca's area. In 1953, Vissarion Shebalin, a Russian composer, had a stroke leaving him with expressive and receptive dysphasia; he could neither understand speech nor talk but remained able to write music. We now know that some shared aspects of music and language are processed together; one example is syntax. For language, syntax is the combination of words required to generate a meaningful sentence; for music, it is the combination of notes required to make a meaningful composition. Brain scan studies seem to show that a part of the frontal lobe is used both to construct meaningful language and for musical "sentences".

hearing music – songs without words

The auditory cortex breaks up the components of sounds, coming in via the thalamus, into tone, melody, rhythm, harmony and timbre. Each component is dealt with by a different part of the brain but, in general, the right temporal lobe is used most, particularly in non-musicians. Tone is processed largely by the auditory cortex itself, while the right temporal lobe analyses melody. Both temporal lobes are needed to determine the key. Harmony and timbre are analysed by the right temporal lobe and rhythm by either, depending on the duration of the sound. Clicking, speech-like sounds are dealt with by the left temporal lobe, while longer rhythmical sounds are processed by the right. Even imagining music lights up these areas of the brain.

Musicians have a larger part of the brain devoted to hearing and a greater response to sounds in general. In fact, the auditory cortex in musicians is about 1.3 times as large as in non-musicians. Similar changes occur in the sensory cortex, motor cortex, cerebellum and corpus callosum. Although these diverse areas are all used in processing music, the most important component of music is the emotion it conveys.

MOTHERESE – THE UNIVERSAL, INSTINCTIVE LANGUAGE OF MOTHERS

In all cultures, parents use "motherese" to communicate with their infants. Motherese is musical in nature and consists of sounds with a wide range of pitch, as well as melodic and rhythmical phrases. Infants react instinctively to this and other musical sounds. They can tell the difference between two similar musical tones, changes in tempo and rhythm and can identify the same melody in different keys. Scientists, led by Peter Hepper in Belfast, Northern Ireland, showed that even fetuses could distinguish the theme tune to the soap opera *Neighbours*, which had been heard every day by their mothers during the pregnancy, from other tunes.

emotions and music

Music directly affects the limbic system, where the brain processes emotion. Brain scans have shown that clashing dissonant sounds light up the regions of the limbic system dealing with unpleasant feelings, while harmonic, pleasing sounds cause a reaction in the regions dealing with pleasurable emotions. So, music directly affects our heart rate, sweating, breathing and blood pressure. Conscious understanding of music is not necessary for its emotional impact, as was shown by a woman with damage to the auditory cortex of both temporal lobes, who was unable to identify any distinguishing features of music (although her speech centres were unaffected). She could not distinguish two different tunes, regardless of how different they were, nor could she recognize music she had heard before. Yet, despite being unable to make sense of music, her emotional reaction to it was completely normal. In other words, although the temporal lobes are needed to understand tunes, the emotional reaction to music is a direct response of the limbic system.

As we shall see in the next chapter, the limbic system is an ancient part of the brain. How can music be directly wired into such a system? The songs of humpback whales and birds share many features with our own music, including similar proportions of percussive and pure tones and similar scales. Whalesong also has a similar length to human music, rarely being shorter than a ballad or longer than a symphony and a similar overall structure, typically following an ABA pattern – theme, elaboration and a return to the original theme, with slight modifications. Therefore it seems likely that music is an innate part of animal communication and that we have simply taken it a step further, as we have with language.

Language, hearing and music are such an important part of human culture and society that it is impossible to imagine the world without them. Vibrations made in the air are captured and analysed by the ear, signals sent to the language area of the brain and converted to symbolic or emotional meaning, within seconds and without effort. This then alters our behaviour, making us make similar vibrations or act in a particular way. Sound interpretation is a wonderful feat of the human brain.

emotions and the limbic system – the heart and the head

I pay no attention whatever to anybody's praise or blame. I simply follow my own feelings.

Wolfgang Amadeus Mozart (composer and musician, 1756–1791)

The best and most beautiful things in the world cannot be seen or even touched. They must be felt within the heart.

Helen Keller (author, 1880–1968)

history – from smells to emotions and memories

The *limbic system* circles the inside of the hemispheres, lying largely on the inside of the temporal lobes. The main components are the hippocampus, the amygdyla and the hypothalamus. In 1878, the French surgeon Broca noticed that the circling structure, or something similar, was present in the brains of all mammals and he deduced that this part of the brain was involved in "animalistic" processes, as opposed to the intellectual functions of the rest of the cortex. This part of the brain also deals with the sense of smell (until very recently, the entire limbic system was called the *rhinencephalon*, or "nose brain"). This association with the sense of smell did the limbic system no favours, because the sense of smell was not regarded as of great importance by the scientific or medical community. As a result, the significance of the limbic system has only just begun to be understood.

Not everyone thought the limbic system just dealt with the sense of smell. One idea was that it acted as a bridge between what we expect and what we actually experience and therefore could be important for

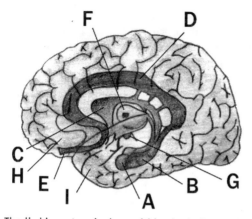

Figure 17 The limbic system is deep within the brain, at the interface of the hemispheres. Only one side is shown here for clarity. A: amygdala, B: hippocampus, C: fornix, D: cingulate gyrus, E: olfactory nerves, F: thalamus, G: mamillary bodies, H: orbitofrontal cortex, I: temporal pole.

memory. Another was that it might play a role in "reactions associated with the expression of emotion". Another was that it might control general mood.

The hints that the limbic system might have a serious role to play in human behaviour were brought together in the late 1930s, when the neurologist, Papez, published his paper 'A Proposed Mechanism of Emotion', using examples of people with problems with the limbic system. He proposed that emotion was built up in the hippocampus, transferred to the mammillary bodies and from there relayed, via the thalamus, to the innermost layer of cortex, the *cingulate gyrus*, where it could be experienced. These ideas were modified a little over the years to include other parts of the brain, but, in general, the idea that the limbic system deals with emotion, memory and sense of smell is what we believe today. In honour of his work, this pathway is known as the "Papez circuit".

evolution of the limbic system – bringing out the animal in us all

The Papez circuit is just part of the limbic system, which also includes the amygdala and the front parts of the brain. The limbic system can be loosely divided into three sections, each with its own function.

First is the amygdala and the front part of the hippocampus. They are primarily involved in self-preservation, particularly feeding, the search for food and the fighting and defensive actions required to eat and obtain a meal. In most animals the olfactory system is a major input into this part of the limbic system, as is the hypothalamus, which signals hunger and satiety. In particular, the amygdala is responsible for learnt fear responses, preparing the body for fight or flight. The fear response from the amygdala is based on recognizing a fearful situation, which may be quite complex. This means the input has already been through most of the brain, to be interpreted and assessed for threat, before it arrives at the amygdala.

Second are the collections of neurons just in front of the thalamus, (the *septal nuclei*), the inner part of the cingulate gyrus and the tail end of the hippocampus. This part of the limbic system deals with reproduction and is involved in sexual function and behaviours, encouraging sociability and mating. The olfactory system and hypothalamus also feed strongly into this section of the limbic system.

Third are the outer parts of the cingulate gyrus and its connections with the outer layers of the brain. The front parts of the thalamus feed into this section, whereas the olfactory system and hypothalamus do not. This part of the limbic system is highly developed in mammals, but does not exist at all in reptiles. It is responsible for the parental care and family groups characteristic of mammals. There are connections between this part of the limbic system and the pain centres, suggesting that the pain of family separation is hard-wired into us.

some of the emotional circuits – how Vulcans miss out

the amygdala

The amygdala receives simple threat signals, such as the sound of an explosion, direct from the thalamus. No complicated processing is required to recognize these likely threats, and so they can go directly to the amygdala. If the threat signal is more complicated, for example, that represented by a particular face or social situation, the signal has to come from an association centre in the outer cortex, because it has to have been recognized as a possible threat. The outputs of the amygdala feed directly into the hypothalamus and brainstem, where they activate the appropriate responses, such as increased heart rate, sweating and breathing. There is also an output to the cingulate gyrus, which processes the emotional significance of the signal. Without this circuit, we would be unable to distinguish the emotional difference between a child with a toy gun and an adult with a real one. During surgery, it is possible to stimulate the amygdala with an electrical probe, which results in people describing anything from mild anxiety to anger, terror and a sense of "someone behind me".

In evolutionary terms, the more social the species, the larger the amygdala. Humans with damage to the amygdala lose the ability to interact appropriately with others. Some people develop hypersexuality and, just as an infant does, become fixated on putting things in their mouth. In general, amygdala damage leads to placidity, muted emotions and an inability to recognize aggression.

The amygdala also has a critical role in memory. In a right-hander, the right amygdala is used for subconscious emotional learning while the left is used for conscious emotional learning.

the temporal pole

The temporal pole is the tip of the temporal lobe and receives inputs from most other areas of the limbic system. A right-handed person uses the left temporal pole to remember the name that goes with a face. The full role of the right temporal pole is not clear but we can use it to recognize a sad face.

the cingulate gyrus

While the amygdala recognizes threats and controls temper, the cingulate gyrus assesses the emotional significance of the experience, for example, to register the emotion behind facial expressions; therefore it is very important for social interaction. People with damage to the cingulate gyrus cannot feel the emotional component of pain and it no longer makes them angry or tearful. This loss of emotion also means that they are no longer driven to pursue activities that they used to enjoy.

The front part of the cingulate gyrus is responsible for "gut feelings". It helps direct our attention towards a potential solution when we are faced with a number of choices, for example in social or novel situations and especially when the possible outcome cannot be known.

the orbitofrontal cortex

The orbitofrontal cortex is just behind the eyes. It plays an important role in matching up the inputs for smell, taste and sight and it is likely to be where we experience "love at first sight". It is extremely important for social interaction and controls responses to social situations, producing changes in heart rate and blood pressure, breathing responses, facial flushing, pupil size, and the "butterflies". Learning to react to and control these autonomic functions is a great part of adolescence.

This part of the cortex acts as our conscience, helping us to censor, self-monitor and incorporate experience into decisions about behaviour. We use it to guess what mental state others are in, allowing us to reason out social problems – we see things from their point of view. One could argue that it most defines what it is be a social animal and is the part that seems to be underfunctioning in forms of autism such as Asperger's syndrome.

The cingulate gyrus and the orbitofrontal cortex together integrate the intellectual, emotional and autonomic understanding of a situation. If they are disconnected, decisions can be made intellectually, without a full sense of their emotional cost. People who do this all the time are described as sociopathic (previously psychopathic) individuals. They appear to act without a conscience, because not only are they unable to empathize with others but they justify decisions purely upon intellectual reasons. On the other hand, disconnection of emotion and reason can be a useful and important feature of social behaviour. It is what leads to the carrying out of acts of great courage, without think-

NEAR-DEATH EXPERIENCE AND RELIGION

In 1975, a physician, Raymond Moody, hit the best-seller lists with a book about survivors of near-death experiences. Nearly all reported similar pleasant experiences, an "out of body" experience, travelling down a tunnel towards light, meeting a being of light or religious figure who helped them to evaluate their lives and finally a decision to return to the material world. Variations on these themes exist; in general, the presence of loved ones and feelings of security, peace and happiness are common. The religious aspects of the experience tend to coincide with the expectations of the individual, so Christians meet Jesus, while Hindus see the messengers of Yamraj coming to take them away. Such experiences can occur to people who believe they are near death when they are not and similar events are reported regularly by military pilots undergoing blackout G-force training. The limbic system seems to be crucial for these near-death experiences, with abnormal firing being the trigger, which can be caused by lack of oxygen or extreme stress.

ing of the consequences. Why one person happens to be a hero and another a criminal remains a deep mystery.

temporal lobe epilepsy – emotions from nowhere

Epilepsy is a spontaneous electrical discharge that can start anywhere in the brain. Because of this, it is useful for "natural" experiments. Asking people with epilepsy to report their experiences during an attack and recording the electrical disturbance itself allow us to discover how each part of the brain produces our experience of the world. From such studies, we have learnt that damage to the limbic system can cause laughing and crying.

A particular form of epilepsy, in which the electrical disturbance begins in the temporal lobes, can be preceded by strong feelings of emotion. People with temporal lobe epilepsy may also experience symptoms associated with other parts of the nervous system, such as sensation, vision or hearing, which has been taken to mean that signals from all these systems feed into the limbic system. So, stimulation of any sense can directly affect our emotions. The limbic system is the clue to the difference between the heart and the head. When we are guided by our intuition, we are led by the limbic system, processing the signals from internal and external senses. When we are guided by our head, we are led by the newest part of the cortex, thinking about these signals.

Temporal lobe seizures can cause religious and mystical experiences. If the electrical disturbance particularly affects the amygdala, the person may experience bizarre imagery of a sexual, dream-like and religious nature, including demons, gods and ghosts. If the rest of the limbic system is particularly affected, the experience is more generally religious and mystical. The writer Dostoyevsky had temporal lobe seizures, often having visions of angels sounding trumpets and the gates of heaven opening. Here is an example of his own description:

> The air was filled with a big noise, and I thought that it engulfed me. I have really touched God . . . "Yes, God exists!" I cried, and I don't remember anything else. "You all, healthy people, can't imagine the happiness which we epileptics feel during the second before our fit . . . I don't know if this felicity lasts for seconds, hours or months, but believe me, I would not exchange it for all the joys that life may bring."

There is usually a pervading emotional flavour to the experience and it may be unpleasant or frightening or difficult to describe in words, pos-

LSD

Lysergic acid diethylamide (LSD), first synthesized by the chemist Hoffman, was quickly found to cause profoundly deep religious experiences in some. This, coupled with the discovery that some alcoholics had spontaneous profound religious experiences that cured them of alcohol abuse, led to its first medicinal use, as a way to medically induce a mystical revelation in the hope of treating alcohol abuse. It was successful but quickly fell into disrepute, because of its use as a street drug. LSD stimulates the limbic system and not only induces religious feelings but also blurs the line between reality and fantasy.

sibly because it is occurring in isolation, without the involvement of the rest of the cortex to put it into an understandable context.

final feelings

Emotions are, in some sense, a primitive part of the brain, because the circuits responsible are found in many other animals, but they are part of what makes us human. The newer parts of the emotional circuits give us feelings of attachment, family, love, selflessness and empathy and an approach to the world that complements the rational, logical approach of the outer, thinking, cortex.

investigating the brain – discovering the diagnosis

from RIP to PET

> The abdomen, the chest, and the brain will forever be shut from the intrusion of the wise and humane surgeon.
>
> Sir John Eric Ericksen (Surgeon-Extraordinary to Queen Victoria)

Until the twentieth century, investigating the workings of the living human brain was an unsatisfactory (and frankly dicey) undertaking. In the late nineteenth century, physicians made great strides in microscopic examination of the brain but this had the disadvantage of only being possible after death. Still, enormous advances were made in the understanding and classification of brain diseases. In the centuries before, investigation of brain function was limited to the effects of various types of head trauma from various types of weapons and the occasional attempt to drill a hole in the skull to release some of the "badness".

It is never easy to judge the true importance of technological developments in the wider context of history when they are witnessed in the present but it is likely that the advances in computer technology which have allowed the development of scanners capable of investigating a living brain will be seen by future generations as a major development. The reduction in size and, more recently, the increase in speed of computer circuitry has permitted the development of new brain imaging techniques. Although these are currently aimed at improving diagnosis, the next fifty years may well see us able to apply therapies non-invasively. This is not so far-fetched when we consider that radiation therapy for certain types of brain tumour can now be guided using a combination of imaging and multiple small beams of X-rays. Each beam is ineffective alone, but together they multiply their effects in the precise area to be treated.

neurophysiological tests – electricity and brainwaves

> If we could look through the skull into the brain of a consciously thinking person, and if the place of optimal excitability were luminous, then we should see, playing over the cerebral surface, a bright spot with fantastic, waving borders constantly fluctuating in size and form, surrounded by a darkness more or less deep, covering the rest of the hemisphere.
>
> Ivan Petrovich Pavlov (Russian physiologist, 1849–1936)

Pavlov, famous for his research into the "conditioned reflex", in which dogs associated various stimuli with food and began to salivate in response, was presciently describing the sort of electrical activity that we now know takes place constantly within the active brain. Neurophysiology is the branch of medicine concerned with the stimulation and recording of neural activity. Some decades earlier, the French physiologist Flourens stimulated various parts of the brains of both animals and humans with electricity, to see what happened. Fortunately, these techniques became a little more refined over the next 150 years and did not require exposure of the brain. However, direct stimulation of the brain with electricity in the conscious, though sedated, patient is still carried out during modern-day brain surgery. This is particularly important when the surgery is near to the so-called "eloquent" areas concerned with speech and language. If stimulation of an area causes the patient to lose the ability to speak or make sense, then the surgeon knows to avoid encroaching into this region.

evoked potentials – recording nerve activity

Evoked potential (EP) recording is a bit like testing the resistance across a circuit. The basic principle is to stimulate on one side and record, using sticky electrode pads on the head, on the opposite side. The tests are carried out in three "directions": first, front to back, by stimulating the eyes (with a chequered pattern, for example) and recording the activity over the visual cortex; second, in an approximately horizontal direction across the brain, via the brainstem, by providing a stimulus to the ear and recording activity from the auditory nerve; and, finally, by stimulating the nerves in the limbs and recording the response from the top of the head. Evoked potential detects delays in the responses of nerves to such stimuli, which can help in the diagnosis of certain inflammatory diseases, such as multiple sclerosis, where the insulating myelin sheath of the nerves is damaged and the speed of nerve conduction becomes markedly slower.

A more recent technique has come from the development of *magnetic stimulation*. Brief currents are induced in the brain which excite nerve cells. Recording the effects of this in a peripheral muscle tells us about the nerve pathways between the brain and the muscle. This method requires a capacitor capable of storing a high voltage, which is then released, allowing current to flow through a coil, creating a magnetic field that passes into the brain and induces an electrical discharge of the cortex. Although initially a diagnostic tool, it is now finding a use as a therapy, particularly in certain mental disorders, where repeated stimulation at lower currents is used. This is an alternative to the more traditional electro-convulsive therapy.

electroencephalography – making waves

Electroencephalography (EEG) is, broadly speaking, the study of "brain waves". In an EEG, about twelve electrodes are placed at key points over the surface of the head, to record nerve activity. Four main types of brain activity can be seen, classified by the frequency of the waves (from slowest to fastest – delta, theta, alpha and beta). Alpha waves are normal and tend to disappear with mental concentration, particularly if vision is involved. Beta waves are smaller and seen frequently in all age groups. Theta and delta are termed "slow waves". Theta waves occur frequently during sleep and are abnormal if present in excess in an awake adult. Delta waves are only normally found in deep sleep and represent the activity of deeper brain structures. Electroencephalography is particularly useful in picking up abnormal brain activity that might be associated with seizure disorders like epilepsy, head injury, brain tumours, infection and inflammation of the brain, chemical disturbances and some sleep disorders. The test can quickly be carried out with the person awake or asleep, or recorded over a longer period as the person goes about their daily routine.

electro-convulsive therapy – repairing a shocking reputation

Electro-convulsive therapy (ECT) is a psychiatric tool, not part of neurophysiology. Mention ECT to most people and they will immediately recall the conman Randle Patrick McMurphy in *One Flew over the*

Cuckoo's Nest. McMurphy seeks institutionalization, to escape regular prison work. In order to subdue his opposition to the dictatorial regime and his perpetual stirring up of the other residents to do the same, he is subjected to severe electric shock therapy and virtually rendered a zombie. In reality, there probably have been cases where ECT has been used in an attempt to subdue troublesome people. In part through a popular association with ideas of state control of free will, ECT has developed a perhaps unreasonably tarnished reputation. In recent years, however, it has had something of a resurgence, not least because it can be very effective in treating depression and certain other mental disorders that are resistant to drug treatments. The effects can be apparent quite quickly, though several courses are the norm. Electro-convulsive therapy seizures usually last about thirty seconds to one minute; the patient is given muscle relaxants and anaesthetic and monitored by an anaesthetist. There are still areas of controversy – for example, whether to apply the electric current to one or both sides of the head – and concerns about short-term memory loss, a recognized complication. However, the loss of quality of life associated with severe mental illness and factors such as suicide risk without treatment must also be considered. As with many therapies, a balance has to be struck between risks and benefits, but ECT is becoming a more acceptable option.

the amytal test – literally half-asleep

For nearly all right-handed and most left-handed people, the left side of the brain is dominant for language skills. Modern brain imaging techniques are beginning to be able to show this, by asking people to perform language tasks inside a scanner, so we can see what parts of the brain are being activated. These methods are still in development, but there is a different approach, which involves anaesthetising half of the brain.

Sodium amobarbital (*amytal*) is popularly known as the "truth drug" (though this can apply to several drugs, all of which induce a state of drowsiness and disinhibition). About fifty years ago, the Japanese neurologist Juhn Wada tried infusing amytal into one carotid artery (the left and right carotid arteries are the main blood supply to their respective halves of the brain), the effect of which was temporarily to anaesthetize that hemisphere. Under this anaesthetic, a neuropsychologist then showed the person pictures and words. After the drug wore off, over the next few minutes, the person tried to recall what was shown. The test was then repeated on the other side. Broadly speaking, failure to recall the objects or words implies that the half of the brain "asleep" at that time is responsible for those functions. This obviously has a large impact on a surgeon's willingness to operate in this hemisphere.

imaging the brain – say "cheese"!

The development of neuroimaging has permitted the detailed study of the brain and can justifiably be said to have revolutionized the study of neurological diseases in the living patient than any other single development. Cerebral imaging involves achieving spatial ("where") and temporal ("when") resolution of "slices" of the tissue being studied. Spatial resolution can be at the tissue level, for example differentiating between grey and white matter (see pages 162–5) or beyond even the microscopic level, for example exploring receptors present on different nerve types (see page 165). Temporal resolution can provide information about blood flow to different parts of the brain and so give some idea of "activation". An extension of this can be to study how the brain uses energy.

X-rays – seeing straight through you

The German physicist Wilhelm Röntgen (1845–1923) was busy passing electric currents through low pressure gases when, in 1895, he discovered that a plate covered with a fluorescent compound appeared to glow in the dark in response to and even at some distance from activity within his sealed cathode ray discharge tube. Briefly immobilizing his wife's hand in the path of these new X-rays (so called because their nature was unknown), in front of a photographic plate, he produced an image of her hand bones, surrounded by flesh, even showing her ring. This first X-ray revealed the principle that different tissues will absorb different amounts of X-rays as they travel through them towards a photographic plate.

The problem with using X-rays to look at the brain is that the skull gets in the way. The bone absorbs a lot of the radiation and the brain itself is rather sensitive to increasing the dose. The images we see on a skull X-ray are pretty useless for looking at brain problems, but in their time they represented a huge advance.

Where X-ray technology retains its use, although magnetic resonance imaging (MRI) is starting to supersede it, is in revealing the blood vessels of the brain – *angiography*. An abnormal expansion of a blood vessel, an aneurysm (see chapter 6) can be made visible by injecting a dye into the blood vessels. The dye is opaque to X-rays and so outlines the aneurysm. In this way, the source of certain types of brain hæmorrhage can be found, with the potential to attach a clip over, or insert a coil within, the aneurysm, to prevent recurrent problems.

pneumo-encephalography and ventriculography – being an air-head

In an account of the history of brain imaging it is important to mention the use of air introduced into the spinal fluid in an attempt to localize intracranial pathology. *Pneumoencephalography* used the fact that replacement of the CSF with air would provide a contrast medium, since the air would be much less dense than CSF to X-rays. Unlike a standard lumbar puncture, this really was a very painful and sometimes fatal experience. Ventriculography involved injecting air directly into the brain cavities, through small holes in the skull. Fortunately all this was superseded in the 1970s, by the invention of the first brain scanner.

computed axial tomography – the CT scan

People still refer to the first brain scanner as the CAT (*computed axial tomography*) scan, though in more recent times the technique has become known simply as CT. Sir Godfrey Newbold Hounsfield (b. 1919), an English electrical engineer, and Allan MacLeod Cormack (1924–1998), a South African physicist, were independently credited with the invention that began a revolution in the study of the brain, and were awarded the Nobel Prize. Hounsfield had already led the design team at EMI Ltd, who, in the late 1950s, created the first all-transistor computer in the UK.

The CT scanner still used X-rays, but the detector was mounted on a rotating frame, with the person's head in the middle of the "dough-nut", so that the absorption of X-rays passing through the head could be measured at a series of angles. The original scanner, first used on a patient in 1972, took several hours to acquire one single slice of brain data and twenty-four hours to reconstruct it into an image. Today, a slice can be obtained in less than a second and reconstructed instantly (due to advances in computing, rather than the mathematics of image reconstruction).

The next development came in the late 1980s, with the invention of spiral CT. In this, the X-ray camera rotates spirally downwards and can encompass data from an entire organ within half a minute. This has also permitted the development of CT angiography.

Computed axial tomography is still restricted, in the field of brain imaging, by the limitations of X-rays and the thick skull bones getting in the way. This is a particular issue when trying to reveal the back of the brain, where the cerebellum and brainstem lie. Although CT will pick up large abnormalities of the brain tissue, such as tumours, the resolution of the scan is not good enough to pick up smaller ones. What CT still does better than any other brain scan is to detect blood in or around the brain soon after a brain haemorrhage (it appears white on the image) and also small fractures and other abnormalities of the skull bones. In these respects, it keeps its place alongside the real revolutionary invention in neuroimaging – MRI.

magnetic resonance imaging – picking up good vibrations

Nuclear magnetic resonance (NMR) forms the basis for *magnetic resonance imaging*. In the 1950s, it was discovered that different materials resonated at different magnetic field strengths, similar to the way the edge of a wine glass can "ring" when vibrated at a particular frequency.

Research into MRI began in the 1970s and the first scanners were tested on patients in 1980. The invention is credited to the British physicist Peter Mansfield (b. 1933) and the American chemist Paul Lauterbur (b. 1929), who were awarded the Nobel prize for their efforts. Another American scientist, Raymond Damadian, looking at tumours in cancer patients, had published ideas about NMR and some credit him with producing the first MRI images of the human body.

How does MRI work? All matter is made of atoms, and all atoms consist of a nucleus and surrounding, negatively charged electrons. The nucleus is made of one or more positively charged protons and may also contain neutrons (with no charge). The simplest atom is hydrogen, which contains a single proton and a single electron. Magnetic resonance imaging detects protons, which means it is good at detecting hydrogen. Because hydrogen is a principal constituent of water (H_2O), which forms seventy per cent of the human body, MRI is particularly good at imaging the human body.

Applying a magnetic field of sufficient strength causes protons to align themselves in the direction of the magnetic field. Current MRI scanners have a magnetic field strength of anywhere from fifteen thousand to ninety thousand times stronger than the pull of gravity on Earth. (This explains why all metal objects have to be removed at the door to an MRI scanner, since they are otherwise likely to fly at incredible speed across the room, stopping at nothing on the way. People with shrapnel injuries, including metal-workers who often have pieces in their eyes, those with some types of metal plates attached to their bones or those with joint replacements have to be carefully screened before they can undergo MRI, for the same reason.) To produce a magnetic field that strong requires super-conductors capable of tolerating the high levels of electrical current involved and an external "jacket", filled with liquid helium, to prevent overheating.

In the person lying inside the scanner, the protons within their brain tissues all line up in the same direction. The person has no awareness of this and does not feel at all peculiar. Then, bursts of radio waves are generated, which causes the protons to resonate and generates the "MR signal". (This causes loud banging noises, which can be frightening for the person undergoing the scan.) When the external radio waves are stopped, the time taken for the protons to line up with the magnetic field again depends on the characteristics of the tissue. The detection part of the scanner picks up the released energy from protons through-out the brain as the radio waves are switched repeatedly on and off and this is reconstructed as a three-dimensional black and white map.

The images produced by a modern MRI scanner look almost as good as if we were to remove the brain and slice it up on the table. There is no X-ray or other harmful radiation involved, which is an additional advantage. Despite this, a significant number of people find that they do not like the enclosed space of MRI scanners – the "torpedo tube" as many refer to it. They may not have suffered from claustrophobia before, but the combination of a small space and loud noise is some-times too much. Fortunately, most brain scanners have an intercom with the outside world for reassurance and even television for distraction.

Is there a limit on the field strength of MRI? Scanners with mag-netic fields of more than 200,000 times the earth's gravitational field now exist, although only in research units. Images from these scanners can show the tiniest blood capillaries within the brain tissue, though the scientists who underwent the scans also experienced mild visual hal-lucinations, presumably through some activation of their visual cortex. Magnetic resonance imaging has continued to develop in other ways. Not only does it provide exquisite pictures of the anatomy of the brain and spinal cord to a degree that CT cannot match, but refinements now mean that the blood vessels of the brain can be seen without the need for injection of any dyeing agent.

Developments in the physics of MRI have opened the gateway to an exciting field – "functional" MRI. With this technique, it is possible to study the flow of and use of oxygen within the blood as it passes through the brain. This can be combined with various tasks, for example looking at pictures that provoke strong emotions, to reveal which parts of the brain are activated. For the first time, scientists are gaining real insights into how the brain functions.

positron emission tomography – brain meets anti-matter

Functional brain imaging was first performed using *positron emission tomography* (PET). This technique uses antimatter, in the form of positively charged electrons, to act as radioactive "tracers" in the blood. The scientist Michael Phelps (b. 1939) is cited as the major contributor to development of the first PET camera, built in 1973 at Washington University, St Louis, USA. The first commercial whole-body PET scanner appeared at the end of 1976.

Positron emission tomography relies on a radioactive phenomenon, "positron decay". Certain radioactive materials will release positively charged particles, called positrons, as they decay. Because a positron is a positively charged electron, it is the "antimatter" counterpart of the normal and abundant negatively charged electrons surrounding all atoms. While we are not quite up to "warp drives" and "transporter beams", there is more than a hint of *Star Trek* about it all. When a positron collides with its "matter" equivalent (a negatively charged electron), energy is released in two beams, at 180 degrees to one another; a ring of detectors around the person can record this and establish where the collision happened. Millions of such collisions happen after the injection of the tracer. The theory is that, if blood containing the tracer is preferentially taken up in one brain area during a specific task, or the tracer itself binds briefly to a receptor on a group of nerve cells, then "hot spots" will show up where relatively more collisions were detected. Thus, a picture of the activity of the brain or the location of nerve receptors is built up, depending on the type of tracer used.

What do we need to make a tracer? A particle accelerator, or cyclotron, bombards chemical elements at high speed, to create radioactive materials that decay by positron emission. Chemists then attach these molecules to the substance that will be traced. For example, since glucose is the major source of energy for the brain, glucose can be used as a tracer to look at brain activity. Currently, PET has a growing application in the detection of certain types of brain tumour and other cancers, but there are relatively few PET centres and in the main it is a research tool only. In this capacity it has an established reputation in neuroscience and has, in particular, led to a greater understanding of Parkinson's disease.

brain investigations

We now have an arsenal of investigative techniques available to us to study the living brain and these are being continually refined. Increasing the resolution of functional scans and generating an accurate three-dimensional EEG are likely to be future investigative possibilities. With the methods we have now, we can discover a great deal about the brain without great risk and largely without great effort.

living for ever – the fountain of youth

Millions long for immortality who don't know what to do with themselves on a rainy Sunday afternoon.

Susan Ertz (*Anger in the Sky*)

brain death – how do we know?

Brain death is the irreversible loss of brain function. Although criteria vary in different countries and there are enormous ethical, legal and religious ramifications, in general brain death is declared when brainstem reflexes, muscle responses and respiratory drive are all absent. Certain states, such as profound hypothermia or the influence of drugs, have to be excluded first.

In certain circumstances, brain death can be very important to establish. The person's heart may still be beating and all the other body organs working. If the person is kept on a ventilator, they could, in theory, remain in this state permanently, with food being supplied directly into the bloodstream. If brain death has occurred, then there will not even be brainstem reflexes, which include pupil reactions to light and the so-called "doll's eye" movements of the eyes, where the eyes appear fixated on a distant object rather than remaining facing forward. Other tests used are the infusion of cold water into the ear canal, to which the eyes should respond with nystagmus, the blink response to touching the cornea, the gag and cough reflexes. Respiratory drive is tested by removing the ventilator tube and monitoring the accumulation of the waste gas carbon dioxide, normally a strong driving force to breathe. Occasionally, specialized tests must be used to confirm brain death, if there is doubt. Electroencephalography is unreliable but techniques such as angiography, evoked potentials and possibly MRI may have a role.

brain repair

Advances in medical technology mean that components of the body can now be repaired or replaced by machines, transplants or enhanced healing processes. Improvements in living conditions, lifestyle and health care mean that more and more people are living longer and longer. As parts fail or wear out, things that would have previously meant disability or death no longer do for many people. The limiting factor for meaningful immortality will soon be failure of the brain and nervous system. If we could prevent or repair such failure, or circumvent it in other ways, long, high-quality life would be possible. This does not necessarily mean it would be desirable but the ethical implications of that would provide material for a whole book on their own.

Body components belong to one of two types. Organs or limbs, such as the heart, liver or arms, could be removed and replaced with a new "plug-in" version whereas systems such as the network of blood vessels could be tweaked but probably not easily substituted.

Although transplantation is now widely accepted as ethical and desirable, there was widespread opposition when it was first proposed. For some religious groups this objection still exists. Why? Transplants involve the assumption that we are not changing the identity of the subject. What would ancient peoples have thought of a heart transplant? For some, it would have been akin to an emotion or soul transplant and the person would not be the same after the procedure as before. This same problem confronts us when we contemplate procedures involving the brain.

mending the brain – changing ourselves

If forced to choose, most people would identify the brain as the defining component of self. Thus, a head transplant from Andy to Bob would really be a body transplant from Bob to Andy. When we talk about "brain repair" we are therefore talking about changes or transplantation of small parts of a brain, because anything more is changing the person's identity. We now need to consider the circumstances in which this becomes important.

Age-related brain changes are of several different types. Neurons daily die in huge numbers from the day we are born to the day we die. We lose about a neuron a second from the cerebral cortex, or thirty-one million a year. After the first few years of life, no more new neurons are made but because there are a hundred billion of them, the loss of a few million a year makes little difference. By the age of seventy, the average person still has ninety-seven per cent of the nerve cells they had at twenty-one. But, as we have seen, in the brain, location is everything. If the nerve loss is concentrated in an important area, it may give rise to symptoms once a critical level is reached. This could happen in some people sooner than others, because they had fewer neurons to start with or were affected by a disease that damaged the neurons already there. Another possibility is that this rate of nerve cell loss is programmed genetically. This sort of critical neuron loss is known as *neurodegeneration*. Examples of neurodegeneration are Alzheimer's disease, in which there is loss of neurons in the temporal lobes, causing memory loss; Parkinson's disease, which affects part of the movement control circuit (the basal ganglia); and motor neuron disease, which affects nerves used for voluntary muscle control, leading to wasting and paralysis. In these conditions, the strategies for treatment have been either prevention of the degeneration or replacement of the lost cells. The most promising avenue for replacement is *stem cell therapy.*

A stem cell is an early embryonic cell that can become any cell type if exposed to the right triggers. Such cells have been used for years in people with blood diseases like leukaemia, who receive a bone marrow transplant. The bone marrow stem cells are dripped in through a vein in the arm and simply "know" to go to the bone marrow. There, they divide and specialize until all the different blood cell types are replenished. It would be wonderful to have a similar system for the repair of brain diseases. How far are we from this holy grail? Is it even possible to find it?

A man may receive a transplant from a woman and vice versa. All cells contain a full set of genes, which means it is possible to track transplanted cells by looking for genetic markers. Studies have shown that women receiving bone marrow transplants from male donors have nerve cells in the cerebellum containing a Y-chromosome. The only explanation is that a transplanted bone marrow stem cell migrated to the cerebellum, became a nerve cell and integrated into the recipient's nervous system. If we could harness this process, it might be possible to instruct stem cells where to go and what to become, using a mixture of hormones and chemicals, so that they could be injected into a vein and integrate themselves automatically into the nervous system.

genetics – programmed for life . . .

What stops us living for ever? Some people argue that we have to age and die to make way for the next generation, otherwise the species would not survive. There is no evidence for this and it is not consistent with our current ideas about evolution and selection. It is the gene, not the species, that is being selected by evolution. The individual is a vehicle for transmitting genes to the next generation, as beautifully described by Richard Dawkins in *The Selfish Gene*. As a result, we should be programmed for survival not obsolescence. The fact that we do age means either that ageing is an advantage to the genes (so evolution has designed us to age) or that it is impossible not to age (so there is no solution to the inevitability of ageing). Let us examine each of these in turn.

. . . built for ageing?

Before we go any further, we should define "ageing". The period between conception and adulthood is a time of development, when the number of cells in the body is increasing, size increases and organs mature. Ageing is the process that begins once we reach adulthood. While this is not a precise definition, it at least sets some parameters for the problem.

Is ageing inevitable? When cells divide, they make an identical, or nearly identical, copy of themselves. In theory, this could go on for ever, giving us a never-ending supply of fresh cells to replace damaged or worn out old ones. But cell division has an in-built property that cannot easily be bypassed. In each cell is the master blueprint that contains the coded instructions for making all the components of the cell. This is DNA, a long filament coiled up into chromosomes. When DNA is copied, the enzyme that does the copying reads the filament a little way ahead. It needs to do this to copy the instructions at the present position. This means that when the enzyme reaches the end of the DNA strand, the last few letters of the genetic code cannot be copied, because the enzyme has run out of filament to read ahead from. Unresolved, this would mean DNA strands becoming shorter and shorter as each daughter cell was made. The solution is that the last sections of the DNA strands contain a repeated sequence of the code, which occurs hundreds of times. The sequence is made up of three (of the four) letters of the genetic code, GGGATT. A special enzyme, *telomerase*, adds more of these repeats every time a cell divides. This protects the chromosomes, because even though the copy is shorter than the original, it is only missing a few GGGATT repeats and these can be added automatically by the telomerase. During development, telomerase is active in every cell, but as we start ageing, it is switched off. We know that people who naturally have more GGGATT repeats live longer than those with fewer, so more repeats are better for a longer life. But why don't we just keep

UNDER ATTACK WHILE WE MULTIPLY

The limiting factor for the lifespan of primitive humans was not predation but infection. Having a good immune system is essential for survival, and evolution has built us a good defence. It is so good that we even (unconsciously, via the sense of smell) select our mate to have as different an immune profile from ourselves as possible, so that our offspring have the benefit of both immune systems. But having a good immune system comes at a price. When a baby is developing, it is an invader in the mother's body. Although half of it is from the mother and so would not trigger an immune attack, half of it carries the father's proteins, marking it as foreign. Having a ruthless immune system would mean thinking of the baby as an invading organism and trying to destroy it. To avoid this, the mother's immune system is modified during pregnancy. But some interesting research has recently shown that the better someone's immune system, the longer they will tend to live and the fewer children they will tend to have. There is therefore an evolutionary balance between fighting off infection and being fertile.

telomerase switched on? That way chromosomes would last for ever. Unfortunately, cancer cells keep dividing precisely because the telomerase is switched back on – having immortal cells has its disadvantages.

Is it an advantage for our offspring if we age? Infant humans are totally dependent on their parents. Without them, the chances of survival are virtually zero. It is therefore clear that natural selection would lead to genes that mean adults are likely to survive until their children are old enough to fend for themselves. Many studies have also shown that having grandparents is associated with grandchildren having larger family sizes. In other words, natural selection should make us likely to survive until at least our grandchildren are born and able to learn from us. So, it does not seem likely that evolution has designed us to age. In fact, most of human evolution took place while old age, as we mean it, did not exist. It seems likely that we age not because we are supposed to, but because we have no choice.

the real reason we could not live for ever

Evolution has programmed us to live as long as possible, but it is impossible for evolution to design a human that could live for ever. Despite this, we now live until we are seventy, eighty or ninety and increasing numbers of people are reaching their hundredth birthday. This is not that different from "for ever". The real problem that we cannot currently solve and that prevents us lasting even longer than we do, is that while we can replace hearts, lungs, livers and kidneys, we cannot replace the brain. The nerve cells in a ninety-year-old's brain were there ninety years before. Nerve cells are strange and difficult to maintain. They live in a complex network of connections. Even an incredibly small fraction of nerve cells dying each day soon adds up. If the human body could live two hundred years, we would all be demented when we died and probably all have Parkinson's disease and motor neuron disease as well. Diseases that involve neurodegeneration were essentially nonexistent in human evolution, but will become more frequent as we gradually push back the frontiers of survival.

bionics

One possible solution to the problem of the finite (although long) life of neurons is to replace them with electronic devices that can be upgraded, or renewed, easily. This is done to a limited extent, with cochlear implants that help deaf people hear, spinal implants that may allow some movement following a spinal injury and bionic eyes that send a signal down the optic nerve, allowing a limited form of vision for some people. All these devices leave the body vulnerable to infection at the point of contact and there is also the problem of the body's reaction to the device as a foreign body. There is currently no artificial lobe of the brain, or memory expansion, but these are early days and it is possible that the technology will allow a cybernetic solution to the "problem" of ageing.

If the problems of rejection and infection could be overcome, it is likely that a young person, connected to a device that provided new information, would automatically develop the ability to interpret it. This seems likely because we are not completely hard-wired to expect a particular body: people who are colour-blind or have extra digits develop the appropriate wiring to deal with the situation. The brain develops according to what it is given. There may be a general hard-wiring plan, with "mini computers" designed for particular tasks, sensory and motor pathways, and regions of the brain for different functions, but it is very flexible, particularly in the young.

the end?

When the first cell harnessed the power of electricity to talk to another cell, the development of the brain was set in motion. Brains are every-where, in many forms, but all doing the same things: keeping the body alive, assessing the environment and moving the body away from danger and towards its goals. Human brains also spend their time assessing what other brains may be thinking, despite those thoughts being private. We do this remarkably well, even though those same brains are spending their time trying to control what other brains might think about them.

Where will our brains take us? What will they be in millions years more? How will they interact with future technology?

We started this book with the hope of providing a guided tour of what our brains are and what they do. Even though our brains may think they understand themselves, we still know only a little. We are no longer in the safe port of our knowledge: we have left the harbour and are well on the way to the open sea. Over the next few years, advances in neuroscience will reveal far more again than we currently know. Soon perhaps, we will really understand the mysteries of the human brain.

appendices

Eric Chudler's brain facts and figures

Reproduced and modified with permission from http://faculty. washington.edu/chudler/facts.html

These data were obtained from several textbooks. All are estimates and averages. Check literature for appropriate references. All numbers are for humans unless otherwise indicated.

whole brain statistics

average brain weights

Species	Weight (g)	Species	Weight (g)
adult human	1,300–1,400	newborn human	350–400
sperm whale	7,800	fin whale	6,930
elephant	6,000	humpback whale	4,675
grey whale	4,317	killer whale	5,620
bowhead whale	2,738	pilot whale	2,670
bottle-nosed dolphin	1,500–1,600	walrus	1,020–1,126
Homo erectus	850–1,000	camel	762
giraffe	680	hippopotamus	582
leopard seal	542	horse	532

Continued

Species	Weight (g)	Species	Weight (g)
polar bear	498	gorilla	465–540
cow	425–458	chimpanzee	420
orang-utan	370	California sea lion	363
manatee	360	tiger	263.5
lion	240	grizzly bear	234
sheep	140	baboon	137
adult rhesus monkey	90–97	dog (beagle)	72
aardvark	72	beaver	45
great white shark	34	nurse shark	32
cat	30	porcupine	25
squirrel monkey	22	marmot	17
rabbit	10–13	platypus	9
alligator	8.4	squirrel	7.6
opossum	6	flying lemur	6
fairy anteater	4.4	guinea pig	4
ring-necked pheasant	4.0	hedgehog	3.35
tree shrew	3	fairy armadillo	2.5
owl	2.2	grey partridge	1.9
rat	2	hamster	1.4
elephant shrew	1.3	house sparrow	1.0
European quail	0.9	turtle	0.3–0.7
bull frog	0.24	viper	0.1
goldfish	0.097	green lizard	0.08

Sources: Berta, A., et al. 1999. *Marine Mammals. Evolutionary Biology.* San Diego: Academic Press; Blinkov, S.M. and Glezer, I.I. 1968. *The Human Brain in Figures and Tables: A Quantitative Handbook.* New York: Plenum Press; Demski, L.S. and Northcutt, R.G. 1996. "The brain and cranial nerves of the white shark: an evolutionary perspective" in *Great White Sharks. The Biology of Carcharodon carcharias.* San Diego: Academic Press; Mink, J.W., Blumenschine, R.J. and Adams, D.B. 1981. "Ratio of central nervous system to body metabolism in vertebrates: its constancy and functional basis". *Am. J. Physiology,* **241**:R203–R212; Nieuwenhuys, R., Ten Donkelaar, H.J. and Nicholson, C. 1998. *The Central Nervous System of Vertebrates* **3**, Berlin: Springer; Rehkamper, G., Frahm, H.D. and Zilles, K. 1991. "Quantitative development of brain and brain structures in birds (Galliformes and Passeriforms) compared to that in mammals (Insectivores and Primates)". *Brain Beh. Evol.,* **37**:125–143; Ridgway, S.H. and Harrison, S. 1985. *Handbook of Marine Mammals* **3**. London: Academic Press

brain dimensions

Average human % brain of total body weight – 2%
Average human brain width – 140 mm
Average human brain length – 167 mm
Average human brain height – 93 mm
Average number of neurons in human brain – 100 billion
Number of neurons in octopus brain – 300 million
Number of neurons in *Aplysia* nervous system – 18,000–20,000
Number of neurons in each segmental ganglion in the leech – 350
Volume of the brain of a locust – 6 mm^3

---

composition of human brain

% white matter – 60
% cerebral oxygen consumption by white matter – 6
% grey matter – 40
% cerebral oxygen consumption by grey matter – 94

---

more facts and figures (human brain)

Average number of glial cells – 10–50 times the number of neurons
Number of neocortical neurons (females) – 19.3 billion
Number of neocortical neurons (males) – 22.8 billion
Average loss of neocortical neurons – 85,000 per day (~31 million
    per year)
Average rate of loss of neocortical neurons – 1 per second
Average number of neocortical glial cells (young adults) – 39 billion
Average number of neocortical glial cells (older adults) – 36 billion
Length of myelinated nerve fibres in brain – 150,000–180,000 km
Number of synapses in cortex – 150 million
Difference between number of neurons in the right and left
    hemispheres – 186 million (more on left)
Weight of hypothalamus – 4 g

Volume of suprachiasmatic nucleus – 0.3 mm^3
Number of fibres in pyramidal tract above decussation – 1,100,000
Number of fibres in corpus callosum – 250,000,000
Area of the corpus callosum (midsagittal section) – 6.2 cm^2
Time until unconsciousness after loss of blood supply to brain –
    8–10 seconds
Time until reflex loss after loss of blood supply to brain – 40–110
    seconds
Rate of neuron growth (early pregnancy) – 250,000 per minute

brain parts: proportion by volume (%)

	Rat	Human
Cerebral cortex	31	77
Diencephalon	7	4
Midbrain	6	4
Hindbrain	7	2
Cerebellum	10	10
Spinal cord	35	2

Source: *Trends in Neuroscience*, November 1995

composition of brain and muscle

	Skeletal muscle (%)	Whole brain (%)
Water	75	77–78
Lipids	5	10–12
Protein	18–20	8
Carbohydrate	1	1
Soluble organic substances	3–5	2
Inorganic salts	1	1

Source: McIlwain, H. and Bachelard, H.S. 1985. *Biochemistry and the Central Nervous System*. Edinburgh: Churchill Livingstone

total surface area of the cerebral cortex

Humans – 2,500 cm^2
Lesser shrew – 0.8 cm^2
Rat – 6 cm^2
Cat – 83 cm^2
African elephant – 6,300 cm^2
Bottlenosed dolphin – 3,745 cm^2
Pilot whale – 5,800 cm^2
False killer whale – 7,400 cm^2

more human cortex statistics

Total number of neurons in cerebral cortex – 10 billion
Total number of synapses in cerebral cortex – 60 trillion
Total cerebral cortex volume:
   frontal lobe – 41%
   temporal lobe – 22%
   parietal lobe – 19%
   occipital lobe – 18%
Number of cortical layers – 6
Thickness of cerebral cortex – 1.5–4.5 mm

EEG

Beta wave frequency – 13–30 Hz
Alpha wave frequency – 8–13 Hz
Theta wave frequency – 4–7 Hz
Delta wave frequency – 0.5–4 Hz

sleep

World record, time without sleep – 264 hours (11 days) by Randy Gardner in 1965. Note: In *Biopsychology* (J.P.J. Pinel, Boston: Allyn and Bacon, 2000, p. 322), the record for time awake is attributed to Mrs Maureen Weston. She apparently spent 449 hours (18 days, 17 hours) awake in a rocking chair. The *Guinness Book of World Records* (1990) has the record belonging to Robert McDonald, who spent 453 hours, 40 min in a rocking chair.

cerebellum

Length of spiny terminals of a Purkinje cell – 40,700 microns
Number of spines on a Purkinje cell dendritic branchlet – 61,000
Surface area of cerebellar cortex – 50,000 cm^2
Number of Purkinje cells – 15–26 million
Number of synapses made on a Purkinje cell – up to 200,000

Species	Cerebellum weight (g)	Body weight (g)
Mouse	0.09	58
Bat	0.09	30
Flying fox	0.3	130
Pigeon	0.4	500
Guinea pig	0.9	485
Squirrel	1.5	350
Chinchilla	1.7	500
Rabbit	1.9	1,800
Hare	2.3	3,000
Cat	5.3	3,500
Dog	6.0	3,500
Macaque	7.8	6,000
Sheep	21.5	25,000
Bovine	35.7	300,000
Human	142	60,000

Source: Sultan, F. and Braitenberg, V. 1993. "Shapes and sizes of different mammalian cerebella. A study in quantitative comparative neuroanatomy." *J. Hirnforsch*, **34**:79–92

human cerebrospinal fluid

Total volume of CSF – 125–150 ml
Half life of CSF – 3 hours
Daily production of CSF – 400 to 500 ml
Specific gravity of CSF – 1.007
Colour of normal CSF – clear
White blood cell count in CSF – 0–3 per mm^3
Red blood cell count in CSF – 0–5 per mm^3
Normal intracranial pressure – 150–180 mm of water

Composition of serum and cerebrospinal fluid

	CSF	Serum
Water (%)	99	93
Protein (mg/dl)	35	7,000
Glucose (mg/dl)	60	90
Osmolarity (mOsm/l)	295	295
Na (meq/l)	138	138
K (meq/l)	2.8	4.5
Ca (meq/l)	2.1	4.8
Mg (meq/l)	0.3	1.7
Cl (meq/l)	119	102
pH	7.33	7.41

Source: Fishman, R.A. 1980. *Cerebrospinal Fluid in Disease of the Nervous System.* Philadelphia: Saunders

cranial nerves

Number of cranial nerves – 12
I-olfactory
II-optic
Number of fibres in human optic nerve – 1,200,000
Number of fibres in cat optic nerve – 119,000
Number of fibres in albino rat optic nerve – 74,800
Length of optic nerve – 50 mm
III-oculomotor
Number of fibres in oculomotor nerve – 25,000–35,000
IV-trochlear
Number of fibres in trochlear nerve – 2,000–3,500
Number of neurons in nucleus of the trochlear nerve – 2,000–3,500
V-trigeminal
Number of fibres in motor root of trigeminal nerve – 8,100
Number of fibres in sensory root of trigeminal nerve – 140,000
VI-abducens
Number of fibres in abducens nerve (at exit from brain stem) –
   3,700
VII-facial
Number of fibres in facial nerve (at exit from brain stem) –
   9,000–10,000
Length of nucleus of the facial nerve – 2–5.6 mm
Number of neurons in nucleus of the facial nerve – 7,000
VIII-vestibulocochlear
IX-glossopharyngeal
X-vagus
Length of dorsal motor nucleus of the vagus nerve – 10 mm
XI-spinal accessory
XII-hypoglossal
Number of neurons in nucleus of the hypoglossal nerve –
   4,500–7,500
Length of nucleus of the hypoglossal nerve – 10 mm

---

spinal cord

Number of neurons in human spinal cord – 1 billion
Length of human spinal cord – 45 cm (male), 43 cm (female)
Length of human vertebral column – 70 cm
Length of cat spinal cord – 34 cm
Length of rabbit spinal cord – 18 cm
Weight of human spinal cord – 35 g
Weight of rabbit spinal cord – 4 g
Weight of rat spinal cord (400 g body weight) – 0.7 g
Maximal circumference of cervical enlargement – 38 mm
Maximal circumference of lumbar enlargement – 35 mm
31 pairs of spinal nerves
31 spinal cord segments
8 cervical segments
12 thoracic segments
5 lumbar segments
5 sacral segments
1 coccygeal segment

hearing

Surface area of the tympanic membrane – 85 mm^2
Length of the Eustachian tube – 3.5 to 3.9 cm
Number of hair cells in cochlea – 10,000 inner, 30,000 outer hair cells
Number of fibres in auditory nerve – 28,000–30,000
Length of auditory nerve – 2.5 cm
Number of neurons in cochlear nuclei – 8,800
Number of neurons in inferior colliculus – 392,000
Number of neurons in medial geniculate body – 570,000
Number of neurons in auditory cortex – 100,000,000

Hearing ranges

Young adult human – 20 to 20,000 Hz
Elderly human – 50 to 8,000 Hz
Rat – 1,000 to 50,000 Hz
Cat – 100 to 60,000 Hz
Dolphin – 200 to 150,000 Hz
Elephant – 1 to 20,000 Hz
Goldfish – 5 to 2,000 Hz
Noctuid moth – 1,000 to 240,000 Hz
Mouse – 1,000 to 100,000 Hz
Sea lion – 100 to 40,000 Hz
Most sensitive range of human hearing – 1,000–4,000 Hz

Source: *Discover Science Almanac.* 2003. New York: Hyperion

Length of external auditory meatus (ear canal) – 2.7 cm
Diameter of external auditory meatus – 0.7 cm
Weight of malleus – 23 mg
Weight of incus – 25 mg
Weight of stapes – 2–4 mg
Length of cochlea – 35 mm
Width of cochlea – 10 mm
Number of turns in the cochlea – 2.2–2.9
Length of basilar membrane – 25–35 mm
Width of basilar membrane – 150 microns (at base of cochlea)
Auditory pain threshold – 130 db
Threshold for hearing damage – 90 db for an extended period of
    time

taste

Total number of human taste buds (tongue, palate, cheeks) – 10,000
Number of taste buds on the tongue – 9,000
Height of taste bud – 50–100 microns
Diameter of taste bud – 30–60 microns
Number of receptors on each taste bud – 50–150
Diameter of taste receptor – 10 microns
Diameter of taste fibre – less than 4 microns
Taste threshold for quinine sulfate – 3.376 mg/litre water

smell

Number of human olfactory receptor cells – 12 million
Number of rabbit olfactory receptor cells – 100 million
Number of dog olfactory receptor cells – 1 billion
Number of bloodhound olfactory receptor cells – 4 billion
Surface area of olfactory epithelium (contains olfactory receptor cells) in humans – 10 cm^2
Surface area of bloodhound olfactory epithelium – 360 cm^2
Area of olfactory epithelium in some dogs – 170 cm^2
Area of olfactory epithelium in cats – 21 cm^2
Thickness of olfactory epithelium mucous layer – 20–50 microns
Diameter of olfactory receptor axons – 0.1–0.2 micron
Diameter of distal end olfactory receptor cell – 1 micron
Diameter of olfactory receptor cell – 40–50 microns
Number of cilia per olfactory receptor cell – 10–30
Length of cilia on olfactory receptor cell – 100–150 microns
Concentration for detection threshold of musk – 0.00004 mg/litre air

vision

Length of eyeball – 24.5 mm
Volume of eyeball – 5.5 cm^3
Weight of eyeball – 7.5 g
Average time between blinks – 2.8 seconds
Average duration of a single blink – 0.1–0.4 second
Thickness of cornea – 0.54 mm in centre; 0.65 mm in periphery
Diameter of cornea – 11.5 mm
Thickness of lens – 4 mm
Diameter of lens – 9 mm
Composition of lens – 65% water, 35% protein
Number of retinal receptor cells – 5–6 million cones; 120–140 million rods
Number of retinal ganglion cells – 800,000–1,000,000
Number of fibres in optic nerve – 1,200,000
Number of neurons in lateral geniculate body – 570,000
Number of cells in visual cortex – 538,000,000
Wavelength of visible light (human) – 400–700 nm
Amount of light necessary to excite a rod – 1 photon
Amount of light necessary to excite a cone – 100 photons
Location of the greatest density of rods – 20° from fovea
Highest density of rods – 160,000 per mm^2
Peak density of rods (cat) – 400,000 per mm^2
Density of cones in fovea – 200,000 per mm^2
Diameter of fovea – 1.5 mm
Intraocular pressure – 10–20 mm Hg
Volume of orbit – 30 ml
Area of retina – 2,500 mm^2
Thickness of retina – 120 microns (range 100–230 microns)
Production rate of aqueous humour – 2 µl/min
Turnover of aqueous humour – 15 times/day
% volume of eye occupied by the vitreous humour – 80
Maximal sensitivity of red cones – 570 nm
Maximal sensitivity of green cones – 540 nm
Maximal sensitivity of blue cones – 440 nm

touch

Weight of skin (adult human) – 4.1 kg
Surface area of skin (adult human) – ~1.9 m²
Number of tactile receptors in the hand – 17,000
Number of nerve endings in hand – 200 per cm²
Von Frey threshold (face) – 5 mg
Two-point threshold (finger) – 2–3 mm
Length of Meissner corpuscle – 90–120 microns
Density of receptors on finger tips – 2,500 per cm²
Density of Meissner's corpuscles on finger tips – 1,500 per cm²
Density of Merkel's cells on finger tips – 750 per cm²
Density of Pacinian corpuscles on finger tips – 75 per cm²
Density of Ruffini's corpuscles on finger tips – 75 per cm²
Thermal pain threshold – 45°C

neurons

Mass of a large sensory neuron – 10^{-6} gram
Number of synapses for a "typical" neuron – 1,000–10,000
Diameter of neuron – 4 microns (granule cell) to 100 microns
   (motor neuron in cord)
Diameter of neuron nucleus – 3–18 microns
Length of giraffe primary afferent axon (from toe to neck) – 5 m
Resting potential of squid giant axon – –70 mV
Conduction velocity of action potential – 0.6–120 m/second
Single sodium pump maximum transport rate – 200 Na ions/second;
   130 K ions/second
Typical number of sodium pumps – 1000 pumps/micron² of
   membrane surface
Total number of sodium pumps for a small neuron – 1 million
Density of sodium channels (squid giant axon) – 300 per micron²
Number of voltage-gated sodium channels at each node –
   1,000–2,000 per micron²

Continued

Number of voltage-gated sodium channels between nodes – 25 per micron^2

Number of voltage-gated sodium channels in unmyelinated axon – 100–200 per micron^2

Diameter of microtubule – 20 nm

Diameter of microfilament – 5 nm

Diameter of neurofilament – 10 nm

Thickness of neuronal membrane – 5 nm

Thickness of squid giant axon membrane – 50–100 Å

Membrane surface area of a typical neuron – 250,000 micron^2

Membrane surface area of 100 billion neurons – 25,000 m^2

Typical synaptic cleft distance – 20–40 nm

% neurons stained by Golgi method – 5

Slow axoplasmic transport rate – 0.2–4 mm/day (actin, tubulin)

Intermediate axoplasmic transport rate – 15–50 mm/day (mitochondrial protein)

Fast axoplasmic transport rate – 200–400 mm/day (peptides, glyolipids)

Number of molecules of neurotransmitter in one synaptic vesicle – 5,000

Diameter of synaptic vesicle – 50 nm (small); 70–200 nm

Diameter of neurofilament – 7–10 nm

Diameter of microtubule – 25 nm

Internodal length – 150–1,500 microns

Composition of myelin – 70–80% lipid; 20–30% protein

	Ion concentration (mM)			
	Squid neuron		Human neuron	
	Intracellular	Extracellular	Intracellular	Extracellular
Potassium	400	20	140	5
Sodium	50	440	5–15	145
Chloride	40–150	560	4–30	110
Calcium	0.0001	10	0.0001	1–2

Source: Purves et al. 1997. *Neuroscience.* Sunderland: Sinauer Associates

blood supply

Brain

Brain utilization of total resting oxygen = 20%
Blood flow from heart to brain = 15–20%
Blood flow through whole brain (adult) = 750 ml/min
Blood flow through whole brain (adult) = 54 ml/100 g/min
Blood flow through whole brain (child) = 105 ml/100 g/min
Oxygen consumption whole brain = 46 cm^3/min
Oxygen consumption whole brain = 3.3 ml/100 g/min

Vessels

Blood flow rate through each internal carotid artery = 180 ml/min
Blood flow rate through basilar artery = 380 ml/min
Diameter of vertebral artery = 2–3 mm
Diameter of common carotid artery (adult) = 6 mm
Diameter of common carotid artery (newborn) = 2.5 mm

references

Bear, M.F., Connors, B.W. and Pradiso, M.A. 2001. *Neuroscience: Exploring the Brain,* 2nd edition. Baltimore: Lippincott Williams and Wilkins

Boron, W.F. and Boulpaep, E.L. 2003. *Medical Physiology. A Cellular and Molecular Approach.* Philadelphia: Saunders

Bradshaw, J. 1992. "Behavioural biology" in C. Thorne, ed., *The Waltham Book of Dog and Cat Behaviour.* Oxford: Pergamon Press

Burrows, M. 1996. *The Neurobiology of the Insect Brain.* Oxford: Oxford University Press

Caviness Jr, et al. 1998. *Cerebral Cortex*, **8**:372–384

Farbman, A.I. 1987. "Taste Bud" in G. Adelman, ed., *Encyclopedia of Neuroscience.* London: Elsevier

Groves and Rebec 1988. *Introduction to Biological Psychology*, 3rd edition. Dubuque: Wm. C. Brown

Guyton, A.C. 1986. *Textbook of Medical Physiology.* London: Saunders

Hille, B. 1984. *Ionic Channels of Excitable Membranes.* Sunderland: Sinauer

Kalat, J.W. 1998. *Biological Psychology*, 6th edition. Sunderland: Sinauer

Koch, C. 1999. *Biophysics of Computation. Information Processing in Single Neurons.* New York: Oxford University Press

Moller, A.R. 2000. *Hearing: Its Physiology and Pathophysiology.* San Diego: Academic Press

Nieuwenhuys, R., Ten Donkelaar, H.J. and Nicholson, C. 1998. *The Central Nervous System of Vertebrates*, **3**. Berlin: Springer

Nolte, J. 1999. *The Human Brain.* London: Mosby

Northern, J.L. and Downs, M.P. 2002. *Hearing in Children*, 5th edition. Philadelphia: Lippincott Williams and Wilkins

Pakkenberg, B. and Gundersen, H.J.G. 1997. "Neocortical neuron number in humans: effect of sex and age". *J. Comp. Neurology*, **384**:312–320

Pakkenberg, B., Pelvig, D., Marner, L., Bundgaard, M.J., Gundersen, H.J.G., Nyengaard, J.R. and Regeur, L. 2003. "Aging and the human neocortex". *Exp. Gerontology*, **38**:95–99

Peters, A. and Jones, E.G. 1984. *Cerebral Cortex*

Ridgway, S.H. 1987. *The Cetacean Central Nervous System.* In G. Adelmon, ed., *Encyclopedia of Neuroscience.* London: Elsevier

Schiffman, H.R. 2001. *Sensation and Perception. An Integrated Approach.* New York: John Wiley and Sons.

Shepherd, G.M. 1998. *The Synaptic Organization of the Brain.*

Shier, D., Butler, J. and Lewis, R. 2004. *Hole's Human Anatomy & Physiology.* Boston: McGraw Hill

Sinclair S. 1985. *How Animals See.* New York: Facts on File

Williams, R.W. and Herrup, K. 1988. *Ann. Review Neuroscience*, **11**:423–453

Willis and Grossman 1981. *Medical Neurobiology.* St Louis: Mosby

glossary

Action potential An electrical disturbance in a cell membrane that travels outwards from its starting position with no loss of power. Usually an action potential travels up or down an axon. Action potentials are depolarizations of the cell membrane and are triggered when an initial depolarization reaches a critical threshold (usually 30 mV).

Afferent The afferent fibres of a nerve take a signal towards the brain.

Amino acid An organic acid forming part of a protein. There are twenty amino acids used by most living organisms. Each amino acid is encoded by a three base sequence of DNA. Some amino acids are encoded by more than one sequence.

Axon The long filament of a neuron. Action potentials arriving at the dendrites travel towards the main cell body. When a sufficient signal is reached, the action potential travels down the axon to the end, where it results in the release of neurotransmitters that trigger an action potential in a muscle or another nerve. Axons of sensory neurons depolarize at the far end and the action potential travels up towards the main cell body. Such cells usually have two axons and the signal then travels from the cell body down the other axon into the central nervous system.

Basal ganglia A collection of nuclei at the base of the brain concerned with movement. The basal ganglia are the extrapyramidal system.

Basilar membrane A component of the inner ear, narrow and stiff at one end, broad and floppy at the other. This gives it the property that vibrations through it will set up a standing wave with its peak at a position along its length dependent on frequency.

Brain A collection of nerve circuits at the head end of an animal, required to generate behaviour appropriate to the environment.

Cell body A nerve cell or neuron has a round part, dendrites and an axon. The round part is the cell body and contains the nucleus.

Cell membrane The semi-liquid wall of a cell, made of phospho-lipids. The cell membrane has the property that it can dissolve both water soluble and fatty substances. It also allows the passage of small molecules through, but not large molecules; this is the basis of osmosis. There are also channels and pumps built into it and recep-tors for transmitter signals from other cells. With these components, the cell membrane controls the electric current and behaviour of the cell.

Central sulcus A groove separating the frontal from the parietal lobes. It runs from the midline towards the ears on each side.

Central nervous system The brain and spinal cord.

Cilia (singular cilium) A filament attached at one end to the outside of the cell membrane, with the other end floating free, outside the cell. Cilia can move, or measure movement. In multicellular organ-isms such as humans, they can be used to move substances past the cell (for example mucus), or to detect movement (for example in the inner ear).

Cochlea The organ of the inner ear. The cochlea contains "hair" cells attached to the basilar membrane (q.v.).

Cone Specialized cell containing a light-sensitive pigment reacting to one of the three primary colours red, green and blue.

Cornea The clear part of the eye.

Corpuscle Literally meaning "little body". Red corpuscles are cells containing haemoglobin for transport of oxygen around the body. Meissner's, Pacinian and Ruffini's corpuscles are specialized sense organs in the skin.

Cortex The outermost layer of the cerebral hemispheres. "Cerebellar cortex" is used to describe the outermost layers of the cerebellum. The cortex is probably where most thinking takes place.

Cytoplasm A carefully controlled solution of salts, within a cell, with organelles suspended in it.

Decussation The crossing over of nerve fibres from one side of the body to the other. Most crossing over takes place in the medulla.

Dendrite Small spiky protuberance from a nerve cell body that connects with other nerve cells.

Depolarize An action potential arriving at a location on a cell membrane changes the electrical charge across the membrane from negative inside ($-70\,mV$) to positive. This electrochemical disturbance in the cell membrane triggers depolarization of nearby membrane and activates various systems inside the cell. Depolarization can also occur as a result of chemical signals received at cell receptors and because of physical deformation of the cell membrane. If enough depolarization occurs, a critical threshold is reached and an action potential is triggered.

DNA Deoxyribonucleic acid encodes genetic information as bases, symbolized by the letters G, A, T and C. Bases are read three at a time by the cell machinery and chains of bases are translated into proteins.

Epithelium A lining of cells on a surface. Examples include the skin and the gut epithelium.

Efferent The efferent fibres of a nerve take signals away from the brain.

Extrapyramidal system The part of the motor system concerned with muscle tone and maintenance of the appropriate posture and state of rest, controlled by the basal ganglia. Damage to the extrapyramidal system results in Parkinson's disease.

Fovea (see *macula*)

Frontal lobes The part of the brain lying in front of the central sulcus. The frontal lobes control eye movements, social behaviour, planning of actions and fine movements. The dominant frontal lobe also controls speech output in conjunction with the dominant temporal lobe.

Ganglion (pl. ganglia) A collection of nerve cell bodies.

Grey matter The name given to collections of nerve cell bodies.

Internodal distance Distance between nodes of Ranvier (q.v.).

Iris The coloured part of the eye. The iris is a ring of muscle. Its central diameter controls the size of the pupil.

Limbic system A system found in all mammals and many higher animals. Translates experience into emotion. In humans, the limbic system is tightly linked with regions for processing smell, memory, sexual arousal, social behaviour, temper and fear.

Lobe A large region of brain, generally performing a single type of function or physically distinct from other brain regions. Examples are the occipital lobes, dealing with vision, and the temporal lobes, dealing with memory, speech and emotion.

Macula Part of the retina with a very high density of cones, allowing us to see with high resolution in our central vision.

Medulla oblongata The hind brain, the lowest part of the brainstem, connecting the rest of the brain to the spinal cord. It contains cranial nerve nuclei largely concerned with swallowing and articulation of speech. It used to be known as the "bulb", hence bulbar palsy is weakness of swallowing or articulation. In some texts, the term "brainstem" is used to refer to the medulla only.

Microtubule Part of the cell skeleton, composed of tubulin, which assembles into long tubes. Tangles of microtubules are seen in Alzheimer's disease. Penrose argues that microtubules have quantum physical properties allowing them to be the substrate for consciousness.

Midbrain Part of the brainstem above the pons and below the thalamus, usually regarded as the highest point of the brainstem. It contains nuclei largely concerned with eye movements, as well as the periaqueductal grey matter, important for pain control. The lower part of the midbrain contains the substantia nigra, part of the extrapyramidal system of movement. Damage to the substantia nigra causes Parkinson's disease. The top of the midbrain shows two pairs of small bumps or swellings called the superior and inferior colliculi. These are concerned with reflexes of eye movements and hearing.

Motor neuron A neuron concerned directly with the process of movement. In humans, upper motor neurons connect the brain to the lower motor neurons. Lower motor neurons connect the upper motor neurons to the muscles.

Neuromuscular junction The synapse between a lower motor neuron and a muscle cell.

Neurofilament An important part of the nerve cell skeleton. All cells have a skeleton. Because of their unusual shape, nerve cells demand a skeleton with particular properties. Neurofilaments have three components – light, medium and heavy – and these assemble into rods 10nm in diameter but the length of the entire neuron. With microtubules, they form a rigid structure and transport organelles and substances up and down the axon. Abnormal accumulations of neurofilaments are seen in motor neuron disease.

Nodes of Ranvier Axons in the central nervous system are wrapped in an insulating sheath of myelin. The nodes of Ranvier are breaks in this sheath and are important because they allow saltatory conduction, thus speeding up the transmission of nerve impulses.

Nucleus 1. A central, roughly spherical membrane inside a cell, containing all the genetic material.
2. A collection of nerve cell bodies embedded within brain tissue, usually in the brainstem. In this sense, a nucleus is identical to a ganglion.

Occipital lobe The part of the cerebral hemispheres at the back of the brain dealing with vision.

Olfactory Relating to sense of smell.

Parietal lobe The part of the cerebral hemispheres behind the central sulcus, above the Sylvian fissure and in front of the occipital lobe. The parietal lobes deal with sensation and orientation in space. With the dominant temporal lobes, the dominant parietal lobe also holds the brain region for understanding speech and for generating the "voice" we use for internal dialogue.

Periaqueductal grey matter Grey matter around the central part of the midbrain. This grey matter contains receptors for opioids and is important for the regulation of pain impulses.

Pons Lies between the medulla and the midbrain and contains nuclei concerned with hearing, movement of muscles in the head and sensation of the face. It also contains nerve fibres travelling along its length between the spinal cord and the brain, and nerve fibres traversing it, largely from the cerebellum, which interact with them.

Pupil The central black hole in the iris that allows light into the eye.

Purkinje cell A component of the cerebellum.

Pyramidal tract The pathway for upper motor neurons connecting the cerebral cortex to the lower motor neurons in the brainstem or spinal cord.

Retina Light-sensitive layer at the back of the eye, containing layers of light-sensitive cells and nerves to enhance the image before it is sent back to the brain via the optic nerves.

Sensory neuron A neuron concerned with the transfer of sensation information from the world to the brain. Sensory neurons connected to the skin have two axons, one reaching to the skin, the other reaching into the spinal cord. Their cell bodies lie outside the spinal cord in ganglia. Each spinal cord segment has a dorsal root ganglion on each side.

Sylvian fissure A large groove separating the temporal lobes from the rest of the brain. If each hemisphere is likened to a boxing glove, the Sylvian fissure is the groove between the thumb and the rest of the glove.

Synapse A connection between nerves or between a nerve and muscle. A synapse is a small gap with receptors on the far side, waiting for the release of a chemical transmitter from the near side. When the receptors are activated by the neurotransmitter, the usual response of the far side is depolarization. When the cell membrane is sufficiently depolarized, an action potential will be triggered which travels down the neuron.

Saltatory conduction Literally "jumping" conduction. Myelinated nerve cells conduct action potentials in this way. When an action potential is triggered, it depolarizes the neighbouring cell membrane, but if this is covered in myelin, the action potential sets up a local current so that the neuronal cell membrane the other side of the myelin segment depolarizes. This means the action potential jumps across the myelin segment and reaches the next bit of neuron much more quickly than if it had been conducted the normal way. A neuron may be ensheathed in many myelin segments and saltatory nerve conduction can be extremely fast. When this process is impaired, we experience problems. If neurons in the central nervous system lose their myelin, this causes multiple sclerosis. If neurons in the peripheral nervous system lose their myelin, this causes a neuropathy.

Temporal lobe The part of the brain below the Sylvian fissure. On the outside, the temporal lobes contain regions for processing speech input (with the dominant parietal lobe) and output (with the dominant frontal lobe). On the inner surface they contain regions for processing memory (hippocampus) and emotion (limbic system).

Thalamus The first part of the forebrain and the highest part of the brain that is not in the hemispheres. The thalamus acts as a sensory relay station, processing sensory information and relaying it to the appropriate part of the cortex.

Tract A bundle of myelinated axons in the central nervous system.

Tympanic membrane The ear drum.

Ventricle A fluid filled cavity within the brain. The are four ventricles in humans: the left and right lateral ventricles (originally numbered 1 and 2), the third, which is a slit between the two thalami in the forebrain, and the fourth, which is a diamond-shaped cavity in the hindbrain. The cavity in the midbrain is the cerebral aqueduct; most medical textbooks do not describe it as a ventricle, probably because it is short and thin.

Voltage-gated ion channel Ions are charged particles that, with a partner, form a salt, for example, sodium chloride. They become independent entities when the salt is dissolved in water. Channels are molecular passageways inserted into a cell membrane, allowing ions and other substances in or out. Gates are channels that are regulated in some way. A voltage-gated channel is controlled by the voltage across the cell membrane, letting more ions through depending on the voltage. The action potential relies on voltage-gated channels.

White matter   Collections of myelinated axons in the central nervous system.

further reading

Bear, M., Connors, B.W. and Paradiso, M.A. 2002. *Neuroscience* (book and CD package). Philadelphia: Lippincott Williams and Wilkins

Dawkins, R. 1989. *The Selfish Gene.* Oxford: Oxford University Press

Donaldson, M. 1986. *Children's Minds.* London: HarperCollins

Dunbar, R., Barrett, L. and Lycett, J. 2005. *Evolutionary Psychology: A Beginner's Guide.* Oxford: Oneworld Publications

God on the Brain (Horizon Documentary) summary and transcript available at: http://www.bbc.co.uk/science/horizon/2003/godon-brain.shtml

Greenfield, S. 1998. *The Human Brain: A Guided Tour.* London: Phoenix

Guttman, B., Griffiths, A., Suzuki D. and Cullis, T. 2002. *Genetics: A Beginner's Guide.* Oxford: Oneworld Publications

Haines, 2003. *Neuroanatomy.* Philadelphia: Lippincott Williams and Wilkins

Langman, J. and Sadler, T.W. (eds). 2000. *Langman's Medical Embryology.* Philadelphia: Lippincott Williams and Wilkins

Milgram, S. 2004. *Obedience to Authority: An Experimental View.* London: HarperCollins

Penrose, R. 1999. *The Emperor's New Mind.* Oxford: Oxford University Press

Young, R.M. 1970. *Mind, Brain and Adaptation in the Nineteenth Century: Cerebral Localization and Its Biological Context from Gall to Ferrier.* Oxford: Oxford University Press

index

Note: page numbers in *italics* refer to figures and tables